Oranges

John McPhee was born in 1931, and educated at Princeton University (where he is now Ferris Professor of Journalism) and Magdalene College, Cambridge. He once worked as a television playwright and as Contributing Editor and Associate Editor for *Time* magazine, and has been a staff writer for the *New Yorker* since 1965. He has published a number of other books, including the recent *The Ransom of Russian Art* (1994), *Irons in the Fire* (1997) and *Annals of the Former World* (1998).

He has received a number of honorary doctorates and several awards, including an Award in Literature from the American Academy of Arts and Letters, Princeton University's Woodrow Wilson Award and the Pulitzer Prize for General Non-Fiction for *Annals of the Former World*. John McPhee is a Member of the American Academy of Arts and Letters and a Fellow of the Geological Society of America.

ORANGES

by John McPhee

PENGUIN BOOKS

for Pryde

PENGUIN BOOKS

Published by the Penguin Group
Penguin Books Ltd, 27 Wrights Lane, London w8 5tz, England
Penguin Putnam Inc., 375 Hudson Street, New York, New York 10014, USA
Penguin Books Australia Ltd, Ringwood, Victoria, Australia
Penguin Books Canada Ltd, 10 Alcorn Avenue, Toronto, Ontario, Canada m4v 3b2
Penguin Books (NZ) Ltd, Private Bag 102902, NSMC, Auckland, New Zealand

Penguin Books Ltd, Registered Offices: Harmondsworth, Middlesex, England

First published in the USA by Farrar, Straus and Giroux 1967
First published in Great Britain with a new preface in Penguin Classics 2000
1 3 5 7 9 10 8 6 4 2

Nearly all of the material in this book originally appeared in the *New Yorker* and was
developed with the editorial counsel of William Shawn and Robert Bingham

The moral right of the author has been asserted

Printed in England by Clays Ltd, St Ives plc

CONTENTS

PREFACE

In Pennsylvania Station, New York City—the old Penn Station, said to have been modeled on the Baths of Caracalla—was a machine that split and squeezed oranges. They rolled down a chute and were pressed against a blade. Then the two halves went in separate directions to be cupped and crunched. The juice fell into a pitcher. You paid dearly for the product.

I was a young commuter, scarcely thirty, on my way to Rockefeller Center from my home in New Jersey, and I stopped at the machine almost every morning. From late autumn and on through winter and spring I noticed a gradual deepening of the color of the expressed juice. December was pale cadmium, April marigold, and June a Persian orange. One day, I happened onto an ad in a magazine, paid for by the Florida Citrus Commission, picturing four oranges that to me looked identical but had varying names: Hamlin, Parson Brown, the Washington Navel Orange, the Late Orange of Valencia. How did they differ from one another? I didn't linger over the question. I had to get to work.

Work in those days—the early nineteen-sixties—was at *Time* magazine, where I spent half my days covering show business and the other half trying to figure out what I could possibly do that might attract the favorable attention of William Shawn, editor of the *New Yorker*. I had enough rejection slips from the New Yorker to paper a large wall but I wasn't getting the message. A promising crack in the door finally developed when I wrote a freelance profile of a Princeton University undergraduate who was exceptionally skilled in basketball. His name was Bill Bradley, and, even then, no one who knew him had any difficulty imagining that he would one day become a Rhodes Scholar (playing basketball for Oxford) and eventually a United States Senator often mentioned prominently for elevation above the Senate. But he was still a college student—hounded by the press after his great successes in basketball—and he needed a place in which to seclude himself so that he could complete in timely fashion his undergraduate thesis.

I, meanwhile, had resigned from *Time* magazine to become a freelance, writing wholly for the *New Yorker*, and I was in search of topics, making lists. I thought of the machine in Penn Station, and the four oranges in the ad. While mentioning a number of story possibilities to Mr Shawn, I uttered the single word "oranges?"

He answered right back. He always answered quickly. It seemed impossible to propose any subject to him that he had not thought about before you had. He kept his writers at the far ends of something like bicycle spokes—all separate, all somehow spinning together and apart, with him in the center—and when he turned down an idea he was usually protecting the interests of some writer whose name would never be mentioned. "No. I'm very sorry. No," he would say typically, his voice so light it fell like mist. "That subject is reserved in a general way for another writer." To my question about oranges, though, he said, "Yes. Oh, my, yes." Grandparents soaked up our daughters, Bradley hid out in our house, and my wife and I left for Florida.

I intended only some hundreds of words, a few pages in the magazine. On the Ridge—the slightly elevated spine of Florida—I began by flying around in a helicopter with a citrus nurseryman and learning the lore of bud unions. Citrus does not come true from seed. If you plant an orange seed, a grapefruit might spring up. If you plant a seed of that grapefruit, you might get a bitter lemon. With a graft, however, what you saw was what you got. Scion and rootstock were joined at the bud union.

I had moved on to growers and pickers—still on course for that short piece—when someone remarked to me that if I was going to write about oranges I should visit the University of Florida's Citrus Experiment Station, in Lake Alfred.

It was late March and the Valencias, in their overlapping cycle, were in fruit and in bloom, a phenomenon of this tree, which blossoms fourteen months before the fruit is picked, with the beautiful result that a Valencia tree in spring is under a snowy veil punctuated by spots of bright orange against an evergreen field of dark leaves. Valencias are half of Florida's annual crop. The university's experiment station was a couple of buff, squarish buildings that stood alone, deep within a Valencian forest. You went up a long lane through the groves to find it. When you stepped out and walked through the door, your short article turned into a book.

Some dozens of people in there had doctorates in oranges. Many of them were wearing white jackets of the sort issued at a general hospital. They were working on citrus metabolism, on post-harvest diseases. In a chamber that functioned much like a heart-lung machine, oranges wired to sensors were breathing in oxygen and exhaling carbon dioxide, as oranges do until they die. They are long off the tree when they stop breathing. Dr. William Grierson, looking lonely, had a knight's broken lance over his door, which he said he had splintered in defense of fresh fruit, most of which was disappearing in a frozen sea of concentrate. In the library of the experiment station were a hundred thousand titles on citrus—scientific papers, mainly, but also six thousand books. They crossed a distinguished spectrum from Philip C. Reece and J. F. L. Childs's *Character Differences in Seedlings of the Persian Lime* to Samuel Tolkowsky's monumental *Hesperides: A History of the Culture and Use of Citrus Fruits* and—from several centuries earlier—Giovanni Battista Ferrari's *Hesperides, or Four Books on the Culture and Use of the Golden Apples*. Some of these I took outside and read under trees. I learned that citrus was born in the Orient, migrated westward with civilization, and traversed North Africa with the rise of Islam. There were no oranges in the Holy Land when Christ was alive. Those oranges on the table in various Last Suppers were oranges of Renaissance Italy. Columbus himself brought the first citrus to the New World.

For more than three decades, people have asked me how it came about that I wrote—as they frequently phrase it—"a whole book on oranges."

That was how.

And Bill Bradley finished his thesis.

ONE

ORANGES

THE custom of drinking orange juice with breakfast is not very widespread, taking the world as a whole, and it is thought by many peoples to be a distinctly American habit. But many Danes drink it regularly with breakfast, and so do Hondurans, Filipinos, Jamaicans, and the wealthier citizens of Trinidad and Tobago. The day is started with orange juice in the Colombian Andes, and, to some extent, in Kuwait. Bolivians don't touch it at breakfast time, but they drink it steadily for the rest of the day. The "play lunch," or morning tea, that Australian children carry with them to school is usually an orange, peeled spirally halfway down, with the peel replaced around the fruit. The child unwinds the peel and holds the orange as if it were an ice-cream cone. People in Nepal almost never peel oranges, preferring to eat them in cut quarters, the way American athletes do. The sour oranges of Afghanistan customarily appear as seasoning agents on Afghan dinner tables. Squeezed over Afghan food, they cut the grease. The Shamouti Orange, of Israel, is seedless and sweet, has a thick skin, and grows in Hadera, Gaza, Tiberias, Jericho, the Jordan Valley, and Jaffa; it is exported from Jaffa, and for that reason is known universally beyond Israel as the Jaffa Orange. The Jaffa Orange is the variety that British people consider superior to all others, possibly because Richard the Lionhearted spent the winter of 1191–92 in the citrus groves of Jaffa. Citrus trees are spread across the North African coast from Alexandria to Tangier, the city whose name was given to tangerines. Oranges tend to become less tart the closer they are grown to the equator, and in Brazil there is one kind of orange that has virtually no acid in it at all. In the principal towns of Trinidad and Tobago, oranges are sold on street corners. The vender cuts them in half and sprinkles salt on them. In Jamaica, people halve oranges, get down on their hands and knees, and clean floors with one half in each hand. Jamaican mechanics use oranges to clear away grease and oil. The blood orange of Spain, its flesh streaked with red, is prized throughout Europe. Blood oranges grow well in Florida, but they frighten American women. Spain has about thirty-five million orange trees, grows six billion oranges a year, and exports more oranges than any other country, including the United States. In the Campania region of Italy, land is scarce; on a typical small patch, set on a steep slope, orange trees are interspersed with olive and walnut trees, grapes are trained to cover trellises overhead, and as many as five different vegetables are grown on the ground below. The over-all effect is that a greengrocer's shop is springing out of the hillside. Italy produces more than four billion oranges a year, but most of its citrus industry is scattered in gardens of one or two acres. A Frenchman sits at the dinner table, and, as the finishing flourish of the meal, slowly and gently disrobes an orange. In France, peeling the fruit is not yet considered an inconvenience. French preferences run to the blood oranges and the Thomson Navels of Spain, and to the thick-skinned, bland *Maltaises*, which the French import not from Malta but from Tunisia. France itself only grows about four hundred thousand oranges each year, almost wholly in the Department of the *Alpes Maritimes*. Sometimes, Europeans eat oranges with knives and forks. On occasion, they serve a dessert orange that has previously been peeled with such extraordinary care that strips of the peel arc outward like the petals of a flower from the separated and reassembled segments in the center. The Swiss sometimes serve oranges under a smothering of sugar and whipped cream; on a hot day in a Swiss garden, orange juice with ice is a luxurious drink. Norwegian children like to remove the top of an orange, make a little hole, push a lump of sugar into it, and then suck out the juice. English children make orange-peel teeth and wedge them over their gums on Halloween. Irish children take oranges to the movies, where they eat them while they watch the show, tossing the peels at each other and at the people on the screen. In Reykjavik, Iceland, in greenhouses that are heated by volcanic springs, orange trees yearly bear fruit. In the New York Botanical Garden, six mature orange trees are growing in the soil of the Bronx. Their trunks are six inches in diameter, and they bear well every year. The oranges are for viewing and are not supposed to be picked. When people walk past them, however, they sometimes find them irresistible.

The first known reference to oranges occurs in the second book of the *Five Classics*, which appeared in China around 500 B.C. and is generally regarded as having been edited by Confucius. The main course of the migration of the fruit—from its origins near the South China Sea, down into the Malay Archipelago, then on four thousand miles of ocean current to the east coast of Africa, across the desert by caravan and into the Mediterranean basin, then over the Atlantic to the American continents—closely and sometimes exactly kept pace with the major journeys of civilization. There were no oranges in the Western Hemisphere before Columbus himself introduced them. It was Pizarro who took them to Peru. The seeds the Spaniards carried came from trees that had entered Spain as a result of the rise of Islam. The development of orange botany owes something to Vasco da Gama and even more to Alexander the Great; oranges had symbolic importance in the paintings of

Renaissance masters; in other times, at least two overwhelming invasions of the Italian peninsula were inspired
by the visions of paradise that oranges engendered in
northern minds. Oranges were once the fruit of the gods,
to whom they were the golden apples of the Hesperides,
which were stolen by Hercules. Then, in successive de-
clensions, oranges became the fruit of emperors and
kings, of the upper prelacy, of the aristocracy, and, by
the eighteenth century, of the rich bourgeoisie. Another
hundred years went by before they came within reach
of the middle classes, and not until early in this century
did they at last become a fruit of the community.
Just after the Second World War, three scientists
working in central Florida surprised themselves with a
simple idea that resulted in the development of commercial orange-juice concentrate. A couple of dozen enormous factories sprang out of the hammocks, and Florida,
which can be counted on in most seasons to produce
about a quarter of all the oranges grown in the world,
was soon putting most of them through the process that
results in small, trim cans, about two inches in diameter
and four inches high, containing orange juice that has
been boiled to high viscosity in a vacuum, separated into
several component parts, reassembled, flavored, and then
frozen solid. People in the United States used to con-
sume more fresh oranges than all other fresh fruits combined, but in less than twenty years the per-capita
consumption has gone down seventy-five per cent, as
appearances of actual oranges in most of the United

States have become steadily less frequent. Fresh, whole, round, orange oranges are hardly extinct, of course, but they have seen better days since they left the garden of the Hesperides.

Fresh oranges have become, in a way, old-fashioned. The frozen product made from them is pure and sweet, with a laboratory-controlled balance between its acids and its sugars; its color and its flavor components are as uniform as science can make them, and a consumer opening the six-ounce can is confident that the drink he is about to reconstitute will taste almost exactly like the juice that he took out of the last can he bought. Fresh orange juice, on the other hand, is probably less consistent in flavor than any other natural or fermented drink, with the possible exception of wine.

The taste and aroma of oranges differ by type, season, county, state, and country, and even as a result of the position of the individual orange in the framework of the tree on which it grew. Ground fruit—the orange that one can reach and pick from the ground—is not as sweet as fruit that grows high on the tree. Outside fruit is sweeter than inside fruit. Oranges grown on the south side of a tree are sweeter than oranges grown on the east or west sides, and oranges grown on the north side are the least sweet of the lot. The quantity of juice in an orange, and even the amount of Vitamin C it contains, will follow the same pattern of variation. Beyond this, there are differentiations of quality inside a single orange. Individual segments vary from one another in their content of acid and sugar. But that is cutting it pretty fine. Orange men, the ones who actually work in the groves, don't discriminate to that extent. When they eat an orange, they snap out the long, thin blades of their fruit knives and peel it down, halfway, from the blossom end, which is always sweeter and juicier than the stem end. They eat the blossom half and throw the rest of the orange away.

An orange grown in Florida usually has a thin and tightly fitting skin, and it is also heavy with juice. Californians say that if you want to eat a Florida orange you have to get into a bathtub first. California oranges are light in weight and have thick skins that break easily and come off in hunks. The flesh inside is marvelously sweet, and the segments almost separate themselves. In Florida, it is said that you can run over a California orange with a ten-ton truck and not even wet the pavement. The differences from which these hyperboles arise will prevail in the two states even if the type of orange is the same. In arid climates, like California's, oranges develop a thick albedo, which is the white part of the skin. Florida is one of the two or three most rained-upon states in the United States. California uses the Colorado River and similarly impressive sources to irrigate its oranges, but of course irrigation can only do so much. The annual difference in rainfall between the Florida and California orange-growing areas is one million one hundred and forty thousand gallons per acre. For years, California was the leading orange state, but Florida surpassed California in 1942, and grows three times as many oranges now. California oranges, for their part, can safely be called three times as beautiful.

The color of an orange has no absolute correlation with the maturity of the flesh and juice inside. An orange can be as sweet and ripe as it will ever be and still glisten like an emerald in the tree. Cold—coolness, rather —is what makes an orange orange. In some parts of the world, the weather never gets cold enough to change the color; in Thailand, for example, an orange is a green fruit, and traveling Thais often blink with wonder at the sight of oranges the color of flame. The ideal nighttime temperature in an orange grove is forty degrees. Some of the most beautiful oranges in the world are grown in Bermuda, where the temperature, night after night, falls consistently to that level. Andrew Marvell's poem wherein the "remote Bermudas ride in the ocean's bosom unespied" was written in the sixteen-fifties, and contains a description, from hearsay, of Bermuda's remarkable oranges, set against their dark foliage like "golden lamps in a green night." Cool air comes down every night into the San Joaquin Valley in California, which is formed by the Coast Range to the west and the Sierra Nevadas to the east. The tops of the Sierras are usually covered with snow, and before dawn the temperature in the valley edges down to the frost point. In such cosmetic surroundings, it is no wonder that growers have heavily implanted the San Joaquin Valley with the Washington Navel Orange, which is the most beautiful orange grown in any quantity in the United States, and is certainly as attractive to the eye as any orange grown in the world. Its color will go to a deep, flaring cadmium orange, and its surface has a suggestion of coarseness, which complements its perfect ellipsoid shape.

Among orange groups, the navel orange is an old one. In his *Hesperides, or Four Books on the Culture and Use of the Golden Apples*, Giovanni Battista Ferrari, a Sienese Jesuit priest of the seventeenth century, described it, saying: "This orange imitates to some extent the fertility of the tree which bears it, in that it struggles, though unsuccessfully, to reproduce the fruit upon itself." It is thus a kind of monster. Just beneath the navel-like opening in the blossom end of each navel orange, there is a small and, more or less, fetal orange, usually having five or six pithy segments. The navel strain that we know now originated in Bahia, Brazil, probably as a bud sport, or mutation, of the Brazilian Selecta Orange. In 1870, an American Presbyterian missionary in Bahia was impressed by the seedlessness and rich flavor of this unusual orange with an umbilicus at its blossom end, and sent twelve nursery-size trees to the United States Department of Agriculture in Washington. The department propagated the trees and sent the progeny to anyone who cared to give them a try. In 1873, Mrs. Luther C. Tibbets, of Riverside, California, wrote for a pair of trees, got them, and planted them in her yard. Mrs. Tibbets' trees caught the attention of her neighbors and, eventually, of the world. From them have descended virtually every navel orange grown anywhere on earth today, including the Carter, the Golden Nugget, the Surprise, the Golden Buckeye, the Robertson, and the Thomson. The patriarchal one should by rights be called the Bahia, but merely because of its brief residence in the District of Columbia it has been known for ninety-six years as the Washington Navel Orange.

In the United States, in a typical year, around twenty-five billion oranges are grown. These include, among others, Maltese Ovals, Pope Summers, Nonpareils, Rubys, Sanford Bloods, Early Oblongs, Magnum Bonums, St. Michaels, Mediterranean Sweets, Lamb Summers, Lue Gim Gongs, Drake Stars, Whites, Whittakers, Weldons, Starks, Osceolas, Majorcas, Homosassas, Enterprises, Arcadias, Circassians, Centennials, Fosters, Dillars, Bessies, and Boones, but not—in all of these cases—in any appreciable quantity. Actually, one variety alone constitutes fully half of the total crop. Originally known in California as the Rivers Late Orange and in Florida as the Hart's Tardiff, it was imported into the United States early in the eighteen-seventies in unlabeled packages from the Thomas Rivers Nursery, of Sawbridge-worth, Hertfordshire. The easygoing Mr. Rivers had not only left off the name of the orange trees; he also failed to note where he had found them. They grew to be big, vigorous trees that bore remarkable quantities of almost seedless fruit containing lots of juice, which had a racy tartness in delicious proportion to its ample sugars. As supposedly different varieties, the trees were already beginning to prosper when an orange grower from Spain, traveling in California, felt suddenly at home in a grove of the so-called Rivers Lates. "That," said the Spanish grower, clearing up all mysteries with one unequivocal remark, "is the Late Orange of Valencia."

Out of the bewildering catalogue of orange varieties and strains, the Valencia has emerged in this century as something close to a universal orange. It is more widely and extensively planted than any other. From Florida and California and Central and South America to South Africa and Australia, Valencias grow in abundance in nearly all the orange centers of the world except Valencia. Having given the world the most remunerative orange yet known, Spain now specializes in its celebrated strains of bloods and navels. Only two per cent of the Spanish crop are Valencias, and perhaps only half of that comes from the groves of Valencia itself; much of the remainder grows in old, untended groves near Seville, where cattle wander through and munch oranges on the trees, on either bank of the Guadalquivir.

The Valencia is a spring and summer orange, and the Washington Navel ripens in the fall and winter. The two varieties overlap twice with perfect timing in California—where, together, they are almost all of the total crop—and the orange industry there never stops. In Florida, the Valencia harvest begins in late March and ends in June, and for about four months there is no picking. Florida grows few navel oranges, somewhat to the state's embarrassment. Florida growers tried hard

enough, some seventy or eighty years ago, but the Washington Navel, in the language of pomology, proved to be
too shy a bearer there. Instead, to meet the fall and winter markets, Florida growers have a number of locally
developed early varieties to choose from, and in the main
they seem to prefer three: the Pineapple Orange, the
Parson Brown, and the Hamlin.

The Pineapple developed in the eighteen-seventies
and was so named because its full, heavy aroma gave
packinghouse employees the feeling that they were
working in Hawaii rather than in Florida. The Pineapple
is fairly seedy, usually containing about a dozen seeds,
but it is rich in flavor, loaded with juice, and pretty to
look at, with its smooth-textured, bright-orange skin and
its slightly elongated shape. The skin is weak, though,
and highly subject to decay. Most oranges, with appro-
priate care, will live about a month after they are picked.
Pineapple Oranges don't have anything like that kind of
stamina. (The Temple Orange and the Murcott Honey
Orange, which are not actually oranges, ripen at the
same time that Pineapples do. They are natural hybrids,
almost certainly tangors—half orange, half tangerine—
and they are so sweet that people on diets sometimes eat
them before dinner in order to throttle their appetites.
Oranges float, but these have so much sugar in them
that if you drop one into a bucket of water it will go
straight to the bottom. Murcotts were named for Charles
Murcott Smith, one of the first men to propagate them.
Advertisements have, from time to time, claimed that

Temple Oranges were native to the Orient and sacred to a little-known sect of the Buddhist faith, and the seeds from which Florida's trees eventually sprang were stolen from a temple against the resistance of guardian priests. Temple Oranges are in fact named for William Chase Temple, who, long ago, was general manager of the Florida Citrus Exchange.)

Parson Nathan L. Brown was a Florida clergyman who grew oranges to supplement his income; the seedy, pebble-skinned orange that now carries his name was discovered in his grove about a hundred years ago. It tends to have pale-yellow flesh and pale-yellow juice, for, in general, the color of orange juice is light among early-season oranges, deeper in mid-season varieties, and deeper still in late ones.

The seedless, smooth-skinned Hamlin, also named for a Florida grove owner, ripens in October, ordinarily about two weeks ahead of the Parson Brown.

Both Hamlins and Parson Browns, when they are harvested, are usually as green as grass. They have to be ripe, because an orange will not continue to ripen after it has been picked. Many other fruits—apples and pears, for example—go on ripening for weeks after they leave the tree. Their flesh contains a great deal of starch, and as they go on breathing (all fruit breathes until it dies, and should be eaten before it is dead), they gradually convert the starch to sugar. When oranges breathe, there is no starch within them to be converted. Whatever sugars, acids, and flavor essences they have were necessarily acquired on the tree. Hence, an advertisement for "tree-ripened" oranges is essentially a canard. There is no other way to ripen oranges. It is against the law to market oranges that are not tree-ripened—that is to say, oranges that are not ripe. Women see a patch or even a hint of green on an orange in a store and they seem to feel that they are making a knowledgeable decision when they avoid it. Some take home a can of concentrated orange juice instead. A good part, if not all, of the juice inside the can may have come from perfectly ripe, bright-green oranges.

Some oranges that become orange while they are still unripe may turn green again as they ripen. When cool nights finally come to Florida, around the first of the year, the Valencia crop is fully developed in size and shape, but it is still three months away from ripeness. Sliced through the middle at that time, a Valencia looks something like a partitioned cupful of rice, and its taste is overpoweringly acid. But in the winter coolness, the exterior surface turns to bright orange, and the Valencia appears to be perfect for picking. Warm nights return, however, during the time of the Valencia harvest. On the trees in late spring, the Valencias turn green again, growing sweeter each day and greener each night.

TWO

ORANGE MEN

AT the beginning of the Valencia harvest in 1965, I drove to Florida, looking forward to its fresh orange juice in much the same frame of mind that I had once been in, in Burgundy, when nearing Beaune and the road that leads through Aloxe-Corton, Nuits-Saint-Georges, Vosne-Romanée, Gevrey-Chambertin, and so on up to Dijon. Several years earlier, I had made visits to orange groves in Spain and California, but this was Florida, with nearly fifty million orange trees, yielding more oranges than Spain, Italy, and Mexico—the second, third, and fourth orange countries—put together. As it happened, I had not been in the state since a time when, as a traveling undergraduate without funds, I had lived for the better part of ten days on roadside orange juice, several bags of tangerines, and three bushels of the Late Oranges of Valencia. I was eager to return to a place where—or so it had once seemed to me—people all but brushed their teeth in fresh orange juice.

About a mile south of the Georgia-Florida line on U.S. 301, on a day as hot as summer although it was in fact the first day of spring, I stopped at the Florida Welcome Station to try some of the free orange juice that was proclaimed on a sign outside. A good-looking redheaded girl handed me a three-ounce cup of reconstituted concentrate. It was good concentrate, as I remember, but I felt a little nonplussed. After driving for some hours, I found myself on Interstate 75 near Leesburg, in Lake County, where the Ridge begins. The Ridge is the Florida Divide, the peninsular watershed, and, to hear Floridians describe it, the world's most stupendous mountain range after the Himalayas and the Andes. Soaring two hundred and forty feet into the sub-tropical sky, the Ridge is difficult to distinguish from the surrounding lowlands, but it differs more in soil condition than in altitude, and citrus trees cover it like a long streamer, sometimes as little as a mile and never more than twenty-five miles wide, running south, from Leesburg to Sebring, for roughly a hundred miles. It is the most intense concentration of citrus in the world. The Ridge alone outproduces Spain and Italy. Its trees, planted in rows determined by surveyors' transits, are so perfectly laid out that their patterns play games with your eyes. The trees are dark and compact, like rows of trimmed giant boxwoods, usually about fifteen or twenty feet high. On the Ridge, as in the Indian River section of eastern Florida, citrus plantations are called groves; in California, they are generally called orchards. Citrus trees are evergreen, and in the ancient world they were coveted for their beauty long before anyone ever thought to eat their fruit. Of all the descriptions of them that I have ever run across, the one I prefer is contained in these three lines by an eighth-century Chinese poet:

> In the full of spring on the banks of a river—
> Two big gardens planted with thousands of orange
> trees.
> Their thick leaves are putting the clouds to shame.

The poet's name was Tu Fu, and he had so much confidence in his writing that he prescribed it as a cure for malaria. Beyond those three lines, I am unfamiliar with Tu Fu's canon. But I believe in him. Or at least I did that morning at the beginning of the Ridge, where the orange trees were shaming the clouds, and the air was sedative with the aroma of blossoms. Valencia trees, unlike all other orange trees, are in bloom and in fruit at the same time. So most of the trees in every direction were white and green and orange all at once.

After leaving Interstate 75, I noticed a sign on a roadside eating place that had originally said, "FRESH ORANGE JUICE"; the word "fresh" had been painted over with white paint but was still showing through. I dropped in briefly at the Florida Citrus Commission, in Lakeland,

and was invited to have a cup of orange juice, which came out of a dispenser in the front lobby. It was concentrate.

While I was there, I heard the story of the day they first demonstrated the fresh-juice machine that was used in the Florida Pavilion at the 1964–65 New York World's Fair. The thing was set up in the Citrus Commission lobby, filled with fresh oranges, and switched on. As it began to split the oranges and squeeze out pitcher after pitcher full of fresh orange juice, word spread through the building, and employees of the Citrus Commission poured out of their offices and into the lobby, where they drank every drop the machine produced. The next day the machine was crated and sent to New York. People at the Citrus Commission were still talking about it more than a year later, and probably they still are. "Drinking that juice was a real novelty," one man told me. "It was a real party. Everybody was smiling."

In Winter Haven, which is on the Ridge and about equidistant from either end, I took a room in a motel on the edge of an orange grove. Next door was a restaurant, with orange trees, full of fruit, spreading over its parking lot. I went in for dinner, and, since I would be staying for some time and this was the only restaurant in the neighborhood, I checked on the possibility of fresh juice for breakfast. There were never any requests for fresh orange juice, the waitress explained, apparently unmindful of the one that had just been made. "Fresh is either too sour or too watery or too something," she said. "Frozen is the same every day. People want to know what they're getting." She seemed to know her business, and I began to sense what turned out to be the truth— that I might as well stop asking for fresh orange juice, because few restaurants in Florida serve it.

At the next table was a couple who overheard my exchange with the waitress and started a conversation with me. They told me that they lived in Plant City, a town about twenty miles away, and that they had an orange grove on their property, with three kinds of oranges, so that ripe fruit was on their trees almost eight months of the year. All year long, they said, they drank concentrate at breakfast. They hadn't made juice from the fruit on their trees for more than ten years.

After dinner, I drove downtown, and in a hardware store I found a plastic orange reamer on a bottom shelf. I bought a knife, too, and went back to the place on the edge of the orange grove. I picked several oranges, squeezed them, and poured the juice into a tall glass. I had what I wanted, but it had been a long day.

The trees in that particular grove were perhaps thirty feet tall. I learned, eventually, that they had been planted in 1885. Because of their great size, obvious age, and relatively metropolitan setting, they are a landmark among citrus men around Winter Haven, who, giving directions, will often say something like, "Turn left at the stop light. You can't miss it. You'll see a grove of big Seedlings on the corner." Seedlings are often spoken of as a distinct variety of orange, but actually the term is a general one, applying to any tree grown entirely from a planted seed. There aren't many fully grown Seedlings anymore, and those that still exist are usually old and quite tall.

Most citrus trees consist of two parts. The upper framework, called the scion, is one kind of citrus, and the roots and trunk, called the rootstock, are another. The place where the two parts come together, a barely discernible horizontal line around the trunk of a mature tree, is called the bud union. Seedling trees take about fifteen years before they start bearing well, and they bristle with ferocious thorns. Budded trees come into bearing in five years and are virtually free of thorns. In Florida, most orange trees have lemon roots. In California, nearly all lemon trees are grown on orange roots. This sort of thing is not unique with citrus. With the stone fruits, there is a certain latitude. Plums can be grown on cherry trees and apricots on peach trees, but a one-to-one relationship like that is only the beginning with citrus. A single citrus tree can be turned into a carnival, with lemons, limes, grapefruit, tangerines, kumquats, and oranges all ripening on its branches at the same time. Trees that are almost completely valueless for their fruit seem to make the most valuable rootstocks. Most of the trees on the Ridge are growing on Rough Lemon—a kind of lemon whose fruit is oversized, lumpy, ninety per cent rind, and all but inedible. As a rootstock, it forages with exceptional vigor and, in comparison with others, puts more fruit on the tree. Bitter Oranges, or Sour Oranges, the kind usually associated with Scottish marmalade and with Seville, make an outstanding stock in certain soils, notably on the banks of the Indian River.

Brochures and booklets distributed by the citrus industry usually say that orange trees live about forty years, but this is probably meant to suggest an amortization span and, whatever the reason, is too conservative. In Europe, one celebrated tree called the Constable lived for four hundred and seventy-three years. This tree, also known as "François I" and "Le Grand Bourbon," was planted as a seed in 1421 by the gardener of the Queen of Navarre. Its profusion of oranges attracted people from all over the kingdom to the Royal Gardens in Pamplona. Both France and Navarre were at the time ruled by Bourbon kings, and the prize tree was at first coveted and then finally inherited by the family's northern branch. At the beginning of the sixteenth century, the Constable of Bourbon had it in his *orangerie* at Chantilly. In 1532, Francis I moved it to his new castle at Fontainebleau. In the late seventeenth century, Louis XIV installed it at Versailles. It died there in 1894, property of the Republic. A good many trees in Florida are still bearing after more than a century. Hence, that eighty-year-old grove in Winter Haven is not particularly ancient or notable, except for the fact that so many experts seem to think—apparently because of the unusual height of the trees—that it is a Seedling grove. The trees are, in fact, Valencias and Pineapples on Rough Lemon.

The man who told me this was in a good position to know, since his family owns the grove and he himself runs one of the state's most successful citrus nurseries. His name is William G. Adams, and when I called his office to introduce myself and ask if I could visit the nursery, he said, "Fine, I'll be right over to pick you up." He happened to be driving a helicopter when he arrived. We flew at five hundred feet over grove after grove, covering some fifteen miles from Winter Haven to his acreage near Haines City. Since conversation could only be shouted, I spent most of the time looking down at the groves. The most striking thing about them from the air, after the unvarying precision of the planting patterns, was the difference between orange and grapefruit trees. On the ground, somehow, it was not nearly so obvious, but, from above, the grapefruit trees were a light meadow green, and the orange trees seemed as dark as pines. Light-green rectangles of grapefruit trees occurred with a kind of random frequency within the wider and darker majority. There seemed to be about six times as many orange trees. The edges of some groves were lined with stacked cordwood, ordinary wood fires being the most rudimentary form of frost protection. We went over a picking crew, with ladders slanting into the trees, field boxes (they are three feet long with a heavy partition across the middle) suggesting bright-orange dominoes, and tractors pulling carts full of oranges out to a high-sided, open semitrailer parked at the end of the rows of trees. The big truck body itself contained a brilliant orange rectangle, composed of fifty thousand oranges, or half a load. The helicopter was yawing and swaying in a gusty head wind, and Adams— a youthful man wearing an open-necked shirt and a fiber hat with madras band—was having trouble keeping it on a true course. The problem didn't seem to bother him. "Isn't this thing great?" he shouted.

"It sure is," I said. "How long have you had it?"

"Almost three months."

"What did you fly before that?"

"Never flown before. There's nothing like it!"

He tilted the thing far over to one side and slid tumultuously downhill, over some wires, and into his nursery. Although it occupies only sixty acres, the nursery is large enough to accommodate an inventory of a million trees, with seedbeds at one end and deliverable trees at the other. Like most other nurserymen, he said, he gets his Rough Lemons and a good many of his Sour Oranges from the Everglades. They grow wild there, and are picked and sold mainly by Indians. There is, of course, no native citrus in Florida, but trees of all varieties have gone wild, their seeds having been dropped by birds, animals, or men. Adams puts, say, a couple of thousand Rough Lemons through an ordinary meat grinder with a coarse blade, and grinds them up—pulp, peels, seeds, and all. He opens a long furrow, eight inches wide and one inch deep, and fills it up with this pale-yellow mash, which produces seedlings. Graduated from the seedbed when they are about six months old, they are set as individuals in rows and are called liners. They are about a foot and a half tall when they are budded, in the following season.

One of Adams' men was putting Hamlin buds on Rough Lemon stock the day I was there. He began by slicing a bud from a twig that had come from a registered budwood tree—of which there are forty-five thousand in groves around Florida, each certified under a state program to be free from serious virus disease and to be a true strain of whatever type of orange, grapefruit, or tangerine it happens to be. Each bud he removed was about an inch long and looked like a little submarine, the conning tower being the eye of the bud, out of which would come the shoot that would develop into the upper trunk and branches of the ultimate tree. A few inches above the ground, he cut a short vertical slit in the bark of a Rough Lemon liner; then he cut a transverse slit at the base of the vertical one, and, lifting the flaps of the wound, set the bud inside. The area was bandaged with plastic tape. In a couple of weeks, Adams said, the new shoot would be starting out of the bud and the tape would be taken off. To force the growth of the new shoot, a large area of the bark of the Rough Lemon would be shaved off above the bud union. Two months after that, the upper trunk, branches, and leaves of the young Rough Lemon tree would be cut off altogether, leaving only a three-inch stub coming out of the earth, thick as a cigar, with a small shoot and a leaf or two of the Hamlin flippantly protruding near the top.

Another year goes by before the scion is, in the language of the business, grove-ready. Adams and I walked down to the far end of the nursery to inspect some of his grove-ready trees. On the way, he told me that he was born in Lakeland and had studied aeronautical engineering at Georgia Tech, but, on coming out of the Navy in 1946, he had decided to go into the citrus business. He started his own nursery in a corner of a grove his father owned in Alturas, and moved it to Haines City in 1960. The bud unions of his grove-ready trees resembled knees, badly crippled. In a few years, the widening trunks would smooth them out. Meanwhile, Adams would sell the trees for something close to two dollars apiece, planted. He guessed that he would sell two hundred thousand by the end of the year.

At Lake Alfred, about five miles north of Winter Haven, the University of Florida maintains a Citrus Experiment Station, which, with the important parallel efforts of a substation in eastern Florida, of other pomologists at the university itself (in Gainesville), and of the much smaller United States Horticultural Station in Orlando, offers a kind of intelligence service for the citrus industry. The Experiment Station has sixty-five scientists, working on everything from flavor and color to post-harvest diseases, mechanized picking, and the bio-chemistry of citrus metabolism. When the respiration of an orange is under surveillance there, the orange breathes into an apparatus that resembles a heart-lung machine. In the station compound, which is surrounded by orange groves, are five buildings, including a miniature fresh-fruit packinghouse, a pocket concentrate plant, and a citrus library of a hundred thousand titles, most of which are, of course, reprints of papers from scientific journals. The library has about six thousand books and hundreds of doctoral theses, with titles like "A Phytochemical Study of the Florida Oil of Sweet Orange," "Citrus Entomology in the Middle East," "A Study of Microörganisms that Survive the Process of Canning Frozen Concentrated Orange Juice," "The Orange Peel," "Root Distribution of the Hamlin Orange on Three Major Rootstocks in Arredondo Fine Sand," and "An Observational Study of the Behavioral Characteristics of Customers While Shopping for Fresh Oranges," the last by a fellow who didn't notice anything particularly interesting. I went to the Experiment Station soon after I had arrived in Florida and returned at least a dozen times. Nearly all of the visits I eventually made to specific growers, pickers, packers, and others began as suggestions made by the pomologists at Lake Alfred.

Dr. Herman Reitz, the director of the Experiment Station, is noted in his field for, among other things, research that resulted in a series of papers that detailed the remarkable and consistent differences among oranges grown in various positions on a single tree. My talks with him drifted conversationally. He told me that viruses strike citrus trees much as they do humans—anywhere, any time, without pattern or predictability. Once a virus infects a tree, there is no known cure. Humans themselves, by budding young trees with sick buds in citrus nurseries, have been the chief spreaders of at least three kinds of citrus virus diseases—exocortis, psorosis, and xyloporosis—and for this reason the Florida State Budwood Certification Program was created. The most destructive virus disease of all, though, is tristeza, which is usually spread by aphids. In South America a few years ago, twenty million citrus trees were killed by tristeza. The disease particularly affects trees which grow on Sour Orange rootstock. What kills them is a kind of arterial occlusion, in which food particles, coming up through the roots and trunk, are blocked at the bud union. Growers in Florida don't have to worry about tristeza very much at the moment. Although aphids by the trillion crawl all over their trees, the particular kind that carries the worst form of the disease is not present in Florida—yet.

I told Dr. Reitz I had seen a picture of a man in California picking oranges from a kind of crow's nest on the end of a long mechanical arm, the same sort of thing that telephone linemen sometimes use. "That," he said, "is a seven thousand dollar ladder."

Dr. Reitz spends a fair amount of time answering calls from people seeking his advice. Many of them are distraught, having reached for the phone only when, as a result of bad luck or ignorance, they begin to suspect that they are rapidly losing money. He speaks in a soft, amiable voice. "Well, I don't know what it might be from this distance, but it sounds like poor drainage," he said to a man who called while I was in his office. "Is your drainage all right in your opinion?" Dr. Reitz kept saying "Mmm" and "I see" while the caller talked steadily for five minutes, and I gathered that the poor fellow had no real idea whether his drainage was superb or terrible. In the end, Dr. Reitz said gently, "Well, if you have surface water, if it's standing there in pools, it's a safe assumption that you have water all the way down." Replacing the receiver, he told me that citrus trees can take an awful beating from a great many things, but one thing they can't cope with is too much water. They need plenty of it, but if their roots actually stand in it, the roots will rot and the tree will die.

Remembering an ad I had once seen in a New York paper, I asked him about shippers of fresh oranges who proclaim and promise that the fruit they ship is unsprayed. "Some people are convinced that pesticides and chemical fertilizers are going to kill us all," he said, "and there are some people who capitalize on this idea. Practically everthing that is said about pest control and pesticides is true; it's a matter of putting it into perspective. If people sell unsprayed fruit for a premium, they're not selling anything of any intrinsic value." I asked him if unsprayed fruit, as shipped, would differ in appearance from sprayed fruit. He said he wasn't sure, because, frankly, he couldn't even think of a shipper who went in for it. "But it probably looks like this," he said, grinning, and he opened a book called *Florida Guide to Citrus Insects, Diseases, and Nutritional Disorders in Color* to show me a color plate of three oranges whose skins were eighty per cent purple, and two more oranges whose skins were completely khaki.

Dr. Reitz was born in Belle Plaine, Kansas, and is a graduate of Kansas State University. A surprising number of people involved in important ways with citrus in Florida are from places like Kansas, Minnesota, Illinois, and Pennsylvania. One man at the Experiment Station, whom Dr. Reitz recommended as probably the most knowledgeable person in the state on the subject of the harvesting and handling of fresh fruit, obviously outdistances everyone else in the catalogue of strangers who are contributing to the economy of Florida. His name is William Grierson, and he is an Englishman, a former officer in the Royal Air Force.

Despite the tidal rise of concentrate, Grierson has been trying to keep growers and shippers interested in fresh fruit. In his office, in a frame over the door, he has a small broken lance, which he carved himself from a piece of birch to symbolize his jousts in the name of fresh oranges and the kind of success he feels he has had. "I am the leader of His Majesty's loyal opposition," he told me the moment I walked into his office. "The fresh-fruit trade has been almost completely neglected lately. I believe that if growers continue to neglect the fresh-fruit market, they may find, in the next ten or twenty years, that the market for all forms of orange products has suffered. These canners get a blood lust whenever they see an orange. I, among others, have been out lobbying to try to keep the fresh-fruit part of the industry alive, although it isn't becoming for a scientist to be out lobbying. In 1948, as a result of overplanting, growers were busy cutting down orange trees, grafting over to grapefruit, planting over to avocados, and that sort of thing, but then the miracle came. Oranges have quadrupled. The concentrate boom is the boomiest boom since the Brazilian rubber boom. The normal laws of economics are defied. When a freeze comes along now, it's a fortunate disaster. It halves the crop, triples the price, reduces taxes, and saves gasoline. When things are going wrong, they are at their best—all bolstered by the miracle of concentrate. This industry is now hungry for new miracles, but one miracle per industry is considerably above any normal quota. We cannot always rely on natural disasters to keep down the volume of fruit. Incidentally, after the great freeze of 1962—the worst freeze of this century—the leaves all turned manila and fell to the ground. Much of the fruit fell to the ground, too, but a lot of it still hung eerily, and with a macabre beauty, in the trees. They looked like odd Christmas trees covered with bright-orange balls."

Grierson is a trim and well-groomed man in his late forties, with brown hair and a modest mustache. He is in the habit, rare in Florida, of working right through the lunch hour every day, pausing only, while eating a sand-wich at his desk, to leaf through the telephone directory in search of unusual names, a daily amusement. The day I met him, he was beside himself with pleasure after dis-covering Verbal Funderburk. Grierson talks and works swiftly, a bit nervously. It is easy to imagine him in an office in Curzon Street, but it is fairly startling to listen to his more or less Oxonian inflections while look-ing past his shoulder, out of his office window, into an orange grove. "My father was a British Army veteran who decided to run an apple ranch in Wenatchee, Wash-ington," he said. "He went there from England not long after I was born, but unfortunately his health collapsed, and in 1922, when I was five, I was sent back to England. I eventually returned—this time to Canada—with a scholarship to Ontario Agricultural College, in Guelph, then back to England for the war, then back to teach at the University of British Columbia, where I had fourteen hundred and forty biology students. Quite enough for a young teacher. I moved on to various experiment sta-tions, working on apples, and finally got my Ph.D., from Cornell. After that, I worked as chief of fruit-and-vege-table research in the food section of the Canadian De-fence Research Medical Laboratories, in Toronto. If you want to sample a dog's life, try being a Ph.D. in a medical laboratory. I came here in 1952. Citrus is fascinating, and in Florida you don't have to dress in an ornamental manner." He was wearing a white, short-sleeved shirt, open at the collar. "A citrus fruit is, botanically, a berry," he went on. "Did you know that? A special kind of berry called a hesperidium. Many berry-type fruits are full of seeds, but they don't have to be. Citrus is monoecious—both sexes are in the same blossom—but in some varieties the pollen and the ovules are always imperfect and fertilization seldom occurs. Oranges can set fruit parthenocarpically—that means 'by virgin development'—so they can develop a fruit even if the flower isn't fertilized. The fruit, however, will be essentially seedless. If an orange has five seeds or less, it is called seedless. People write in and complain about the seeds they find in seedless oranges. On the other hand, if an orange likes its own pollen, as the Parson Brown and the Pineapple do, it will have dozens of incestuous seeds. The sex life of citrus is something fantastic."

(Citrus scientists have difficulty finding the property lines between varieties and species and between species and hybrids. One astonishing illustration of this came as the result of an attempt, at the United States Horticultural Station in Orlando, Florida, to grow a virus-free Persian Lime. This is the kind of lime, almost perfectly seedless, that goes into everybody's gin and tonic. About fifteen years ago, many Persian Lime trees in Florida were affected by a virus that was drastically shortening their lives. The most common way to create a virus-free strain of a citrus fruit is to plant a seed, since a parent's virus is not transmitted to its seedlings. Persian Limes contain so few seeds, however, that the researchers—

Philip C. Reece and J. F. L. Childs—cut up eighteen hundred and eighty-five Persian Limes and found no seeds at all. So they went to a concentrate plant and filled two dump trucks with pulp from tens of thousands of Persian Limes which had just been turned into limeade. Picking through it all by hand, they found two hundred and fifty seeds, and planted them. Up from those lime seeds came sweet orange trees, bitter orange trees, grapefruit trees, lemon trees, tangerines, limequats, citrons—and two seedlings which proved to be Persian Limes. Ordinarily, a citrus seed will tend to sprout a high proportion of something called nucellar seedlings, which are asexually produced and always have the exact characteristics of the plant from which the seed came. The seeds of the Persian Limes, however, sent up a high proportion of zygotic seedlings, meaning seedlings which arise from a fertilized egg cell. If zygotic seedlings come from parents which are true species, the seedlings will always quite obviously resemble one or the other parent, or both. If zygotic seedlings come from parents which are hybrids, they can resemble almost any kind of citrus ever known. The Persian Lime itself is probably a natural hybrid. The trees that grew from Reece's and Childs' lime seeds are still young, and they copiously produce their oranges and lemons, grapefruit and tangerines every year. The lemons are a type that are not grown, except perhaps in a laboratory, within three thousand miles of Orlando. However, most pomologists who are familiar with this story think that it has only one truly remarkable aspect.

They think it is fairly phenomenal that, out of two hundred and fifty seeds, Reece and Childs got *two* Persian Limes.)

"Have you ever seen a chimera?" Grierson said. Reaching to a shelf behind him, he picked up an orange, and slowly rotated it in his hand. Its skin was a bright and uniform orange, but as the orange continued to turn, the chimera came into my view—a section of the peel that appeared to correspond exactly to the shape of a segment of the flesh inside, framed from axis to axis by a pair of absolutely straight longitudinal lines and colored dark shining green. "It's a mutation of one carpel, perhaps caused by a cosmic ray, at least that's the classic guess," he went on. "They come in many forms. Last month I saw the reverse of this one—an orange that was almost entirely green, with one sharply defined orange section. I've had an orange in here that was one-quarter immune to everything. The rest of it was sick. I tried to give a completely healthy chimera to the wife of a friend of mine not long ago, but she'd have none of it after I'd told her what it was. All the talk about mutations, you know. There is more total weight of citrus fruit each year than of any other tree fruit in the United States, but very little research, actually. There are citrus experiment stations here and in Riverside, California, and in Weslaco, Texas, but that's about all in this country. More than forty stations are working on apples. The procedure here can be difficult for people who deal with subjective matters like taste and aroma. Before they can publish, they have to prove what they are saying mathematically. The public has very low taste perception, anyway. You must meet Bob Rutledge at the Florida Citrus Mutual. He is a phenomenon. If you come down here, you should meet phenomena. You must meet Mac—Dr. Louis Gardner MacDowell, the patron saint of concentrate. You should meet Ben Hill Griffin, of Frostproof, Florida—probably the last of the great orange barons. Too bad most of the interesting people are dead. E. Bean —that was a famous name in oranges once. In Northern cities, grocers used to put up signs advertising 'E. BEAN'S ORANGES HERE.' Bean designed the orange crate, in 1875. He designed the field box, too, the one used in the groves. It weighs fifteen pounds and holds ninety pounds of fruit, or about two hundred oranges. Try slinging something like that around all day! By and large, we're still tied to his damned field box. Today, most oranges are grown either by corporations or coöperatives. Sunkist, in California, is the largest marketing coöperative in the world, but we have smaller ones on the same order here. You should meet a production manager. One of the best in the state is Art Mathias."

Mathias, production manager for the Haines City Citrus Growers Association, works for two hundred and thirty so-called growers, but he is the actual farmer, and, with his crews, produces about a quarter of a billion oranges a year. As citrus men go, he is atypical in that he

loves to eat citrus fruit. He eats four or five oranges
every day and at least two grapefruit, either chilled or
broiled, the latter with maple syrup and cinnamon on
them, left in the oven until slightly brown on top. He
and his wife keep a big glass jar in the refrigerator filled
with home-cut orange and grapefruit sections.

Mathias was born in Evansville, Indiana, but he is a
thoroughgoing Floridian. His father, once an engineer in
an Indiana brewery, moved to Florida in 1920, when
Mathias was ten months old, and eventually became production manager for the Haines City Citrus Growers
Association. Mathias, who now cruises the groves in an
air-conditioned and radio-equipped Chevrolet, and supervises the efforts of mastodonic spray trucks, remembers riding the groves on horseback with his father in an
era when spray machines were pulled by mules. He majored in citrus at the University of Florida, later joined
Haines City as spray foreman, and replaced his father as
production manager in 1959.

Mathias is a handsome man with a somewhat vestig-
ially athletic build, a gray edge around the temples,
and a soft, easily flowing voice. He has an unpretentious
attitude about his work. "There are fourteen thousand
growers and fourteen thousand ways of raising fruit," he
told me. "We're mighty lucky the citrus tree is so hardy,
or we would have killed the species long ago." The
groves he supervises cover a total of seven thousand
acres scattered over some fifty miles of the Ridge.
Among Haines City's two hundred and thirty growers, a

few own as little as five acres. Most own about twenty acres, and the largest single holding in the coöperative is three hundred acres. Owning a block of orange trees is like owning a block of stock. Many of Florida's fourteen thousand growers have never been to Florida. Thirty per cent of them live outside the state. At the moment, mature groves, with trees ten years old and older, are selling for something like four thousand dollars an acre. Membership in the Haines City Citrus Growers Association costs one dollar. If an owner has bearing trees on his property, the coöperative gives him a drawing account, advancing the costs of production against eventual profits, and sending him a check for the difference at the end of the season. Owners with new groves are sent monthly bills until their trees come into bearing. Mathias likes to plant trees fairly spaciously on a twenty-five-foot square, but the decision is the owner's. The more concentrated the pattern the more money the trees will earn in the short term and the less in the long term. Owners who happen to be in the area participate heartily at monthly board meetings, but they leave Mathias alone—just as big-company stockholders are prone to sound off at annual meetings but seldom make a nuisance of themselves in production plants. Actually, Mathias wishes that more of them would come around and have a look at their groves.

It is difficult, though, to walk in the groves, because you sink to your shins in sand. All of Florida was under water fifty million years ago, and the Ridge is the remains of a string of submarine hills. Clay and limestone are under the sand, but the sand is about twenty feet deep. There are traces of phosphate in it and occasional suggestions of organic material, but it would be a scandalous exaggeration to call it sandy soil. It has the texture and porosity of a beach, and feels exactly the same underfoot. "It's a kind of hydroponic deal," Mathias explained to me during the course of a day that I spent riding around with him from grove to grove. "The sand holds up the trees, and we do the rest." His car pitched and rolled in the soft sand, and once in a while was stopped altogether. When this happened, he deftly rocked it back into action. Only occasionally, he said, does he have to radio for a tractor to pull him free. In varying amounts, he feeds the trees nitrogen, phosphorus, potassium, calcium, magnesium, iron, copper, manganese, zinc, boron, and molybdenum. The trees, in turn, are fed upon, or otherwise attacked, by many kinds of mites, insects, and fungus diseases, against which Mathias's crews use a liquid arsenal ranging from a fine mist of petroleum to parathion, a deadly nerve gas developed by German scientists during the Second World War. Workers have to wear gas masks when they are applying it, but parathion is very volatile; it does its work and vanishes. "Citrus is sprayed very little, compared with other fruits," he said. "A few sprays are toxic, and we have to be careful. Most are non-toxic. The hazard to the consumer is virtually nil. By law, even the peel of an orange has to be safe for eating, and all of the spray residue, if there is any, is actually thrown away with the peel."

During the course of the day, Mathias got out of the car several times to correct the technique of men who were spraying young trees, taking off his glasses, putting on a hat, and walking into the fringes of a wet gale of copper fungicide, zinc, manganese, borax, and Chlorobenzilate. Mathias has to mount campaigns against rust mites, red mites, six-spotted mites, Texas citrus mites, mealy bugs, cottony cushion scales, soft scales, black scales, Florida wax scales, purple scales, Florida red scales, snow scales, yellow scales, dictyospermum scales, chaff scales, white flies, white-fly fungi, aphids, plant bugs, orange-dog caterpillars, Mediterranean fruit flies, melanose, and citrus scab, to name a few. The problem sometimes becomes complex. Rust mite, for example, is a creature that gives an unappealing rusty color to an orange. Rust mite is controlled with sulphur-based miticides. But if the weather happens to be hot after a sulphur spray, oranges will get something called sulphur burn, which looks more or less like the effects of rust mite. Sometimes, Mathias has seventy-two hours to stop an infestation—of aphids, for example—or it is too late. Some foes attack underground, most notably the burrowing nematode, a small worm that is the author of a disease called the spreading decline. The nematode feeds on small roots and increasingly cuts off the food supply of the tree, which dies slowly, from the top down, as more and more skeletal branches appear each year and the amount of fruit steadily decreases. When people in Florida are feeling depressed and miserable with some unspecific malady, they sometimes tell one another that they have the spreading decline. Since no one has yet found a way to kill the nematodes without killing the tree, decline brings economic disaster. Whole groves of affected trees and a surrounding margin of healthy trees often have to be bulldozed into a great pyre and burned; after the land they stood on is fumigated, it must be left empty for three years. As we drove along, Mathias would now and again point to areas full of half-dead trees and say, "Decline." Some were all but leafless, and looked like Northern apple trees in February. Once, we were on a secondary road, moving along between healthy, thick-foliaged orange groves, when perhaps fifty acres of treeless land suddenly came into view, covered with new houses, all of which looked alike. "Decline," Mathias explained.

Lightning kills as many orange trees as any disease. Central inland Florida has more thunderstorms than any other area in the United States, and there is not much on the Ridge for the bolts to strike except the trees. Wind damages oranges. Whipping branches against one another, it can put a minute scratch on a new, tiny orange that will develop into a blemish covering a large area of the mature fruit. It is possible that the gusty blasts of spray machines cause as much wind damage as wind itself. Considering the enemies of citrus as a whole, it is no wonder that the trees people keep in their yards for ornamental effect seemed—to me, at least—to be the sorriest-looking trees in the state. For the most part, the trees of the commercial groves are beautiful things that shine with health. They are pruned with enormous machines that move along the middles—the lanes between rows of trees—and, with big circular saws, trim the trees until they are hedged and topped with such uniformity that the Ridge itself seems to be a macrocosmic reproduction of the gardens at Chenonceaux or Villandry.

Abruptly turning off a road and into a grove, Mathias drove about a hundred yards and stopped at the brink of a conical pit that was at least a hundred feet across and about forty feet deep. Valencia trees around the rim tilted crazily over the edge, part of their roots protruding into the air, bleached white by the sun. Fruit nonetheless hung from the limbs of these trees, and some of it had dropped and piled up far below among new scrub-pine trees. Mathias said that the hole had suddenly appeared one day about four years earlier. Something had obviously released a plug under the sand, which had simply run out, taking part of the grove with it. "These sinkholes develop quite frequently," he said. "A house went into one not long ago. Another time, an entomologist at the Experiment Station lost his back yard." Years ago, before the development of the citrus groves on the Ridge, logging crews working the pines that grew there originally were occasionally surprised by the sudden sensation that the earth was giving way beneath them. They ran for their lives, sometimes shouting, "The Devil's got the roots!" Citrus men would be more likely to call it the work of God, however, because the Ridge's tendency to perforate itself with sinkholes has resulted in considerable benefit to the growers of oranges. The holes are caused by the collapse of the ceilings of underground caverns, made by underground rivers in the soluble limestone. If a fissure develops in the limestone, the sand above pours through it. Or, during a drought, when the level of the water in the caverns drops, the ceilings become unsupported and may give way. The sinkhole Mathias showed me was a small one. Some of the older ones are a mile or so across, and, more often than not, are filled with water. There are about thirty thousand named lakes on the Ridge, most of which began as sinkholes and most of which are almost perfectly round. Towns like Winter Haven and Orlando are polka-dot Venices. The advantage to orange growers is that the lakes tend to modify cold air and protect trees from frost. It is not enough merely to have lakes in the general vicinity, however, and the quality of a grove site will vary according to its position in relation to the nearest lake. Mathias went through a grove and stopped at the edge of a pond called Little Lake Hamilton to show me how true this is. Along the north shore, all the orange trees had been severely damaged in the big freeze of 1962. All their leaves had fallen, the fruit was lost, and the trees had later been radically pruned. Along the south shore, which the north wind had reached after passing over the lake, no trees were damaged and no oranges were lost.

Even more than the proximity of lakes, good air drainage can keep a grove relatively safe from frost. This is one of the advantages of the Ridge as a whole. Cold air runs off it like water off a tent. But within its undulations there are countless pockets and knolls that will make one spot relatively dangerous for citrus trees and another spot, actually only a few yards away, relatively safe. Driving along a highway, Mathias pointed out the robustness of trees growing across the top of a slight rise. The car went down a gentle slope. At the bottom, in a pocket where cold air would collect on a night of frost, no trees at all were planted. Some trees on the edge of the pocket had obviously been damaged. At a time when the temperature on the nearby crests is around thirty-six degrees, Mathias said, the temperature down in the pocket might be twenty-four degrees. What passes for a cold night in Florida seldom goes very far below twenty-eight degrees—the temperature at which oranges freeze—and this is why the net effects of a Florida freeze can turn on sensationally slight differences in the location of trees. Mathias left the highway and drove on level ground through a fine grove of Valencias, heavy with fruit. After several hundred yards, he said, "Now watch this." The car started down a hill —not a steep one, merely a discernible incline. The farther down we went, the more the trees showed the effect of the 1962 freeze—increasingly pruned back and threadbare, with fewer and fewer oranges on them and new shoots, much too young to bear, poking up in confused jumbles from the trunks. Farther downhill, the trees had been killed to the bud union, and at the foot of the slope they had been killed to the ground. They still stood there, leafless and gray. The difference in altitude between the dead trees and the trees full of oranges at the top of the rise was twenty feet, if that.

The history of Florida is measured in freezes. Severe ones, for example, occurred in 1747, 1766, and 1774. The freeze of February, 1835, was probably the worst one in the state's history. But, because more growers were affected, the Great Freeze of 1895 seems to enjoy the same sort of status in Florida that the Blizzard of '88 once held in the North. Temperatures on the Ridge on February 8, 1895, went into the teens for much of the night. It is said that some orange growers, on being told what was happening out in the groves, got up from their dinner tables and left the state. In the morning, it was apparent that the Florida citrus industry had been virtually wiped out. The groves around Keystone City, in Polk County, however, went through the freeze of 1895 without damage. Slightly higher than anything around it and studded with sizable lakes, Keystone City became famous, and people from all over the Ridge came to marvel at this Garden of Eden in the middle of the new wasteland. The citizens of Keystone City changed the name of their town to Frostproof.

The twentieth century has had numerous severe freezes, but no great one until 1962, when, on the night of December 11th, northwesterly winds carried a mass of gelid air down from the Arctic with such speed—a thousand miles a day—that the temperatures were still at the killing level when the cold reached Florida. It lasted four nights, and, at its coldest, was cold enough to split the bark of trees, causing a sound like repeated rifle shots. During much of that great freeze, Mathias told me, a fairly stiff wind was blowing. Driving north through a grove, he pointed out that, seen from the south side, the trees appeared to have been completely unaffected. Then he turned the car around and drove the other way, and the same trees looked ragged, battered, and half alive. The worst night was December 13th, when the temperature in some parts of the Ridge stayed below twenty degrees for as much as four and a half hours, and even Frostproof recorded a minimum for the night of twenty-four degrees. Nearly eight billion oranges were lost in Florida. So many trees were either damaged or killed that the ultimate effect was to cut the state's production in half. Immediate losses of ripe fruit on the trees would have been even greater after the 1962 freeze if there had not been a slow thaw, which enabled citrus men to harvest millions of frozen oranges and rush them to concentrate plants for processing.

If freezing air moves into grove land on a relatively calm winter night with an open sky above, the temperature can be, say, twenty-six degrees at ground level and

thirty-four degrees just above the tops of the trees. Because of this phenomenon, known as temperature inver-
sion, some groves have wind machines—great engines
supported on towers with something like an airplane
propeller mounted on them to mix the cold, low air with
the warmer air above. Wind machines are useless on
windy nights and effective only in relatively mild freezes.
Oil heaters (called smudge pots only by tourists), wood
fires, coke fires, and even burning automobile tires are
also used to fight the cold. Such heat sources set up con-
vection currents that stir together the colder and warmer
layers of air. During freezes, some growers turn on overhead water sprinklers—which exist because there are
short but crucial periods of drought in Florida every
year—and spray their trees until they are coated with
icicles. This dramatic system sometimes works, because
when water freezes it releases heat. Although the temperature of the air may have dropped to twenty-five
degrees, water freezing on the fruit and foliage of an
orange tree will hold at thirty-two degrees as long as
more water continues to be applied. It takes a lot of
nerve to wage a water-spray battle with Arctic air. If
the wind is blowing with any speed, it forces the temperature of the ice down, and the fruit—perhaps the
tree, too—is lost. If the period of deep cold lasts for too
many hours, the weight of ice on the tree becomes so
heavy that whole branches break and crash to the ground.
So a kind of race develops between the accumulating
weight of the ice and the arrival of sunrise and the thaw-

ing temperatures of the day. In 1962, many growers lost this race, but some, although they had applied layer after layer of ice until their trees looked like frozen fountains, were saved by the rising sun, which melted away all the ice, leaving perfect oranges, unaffected, hanging on the trees, worth three times as much as they had been a day or so before.

Four days before the arrival of the 1962 freeze, Florida knew it was coming, having been told so by the Federal-State Frost Warning Service, in Lakeland, on which production managers, like Mathias, and every kind of grower, large and small, depend, day in and day out, each winter. A great deal of planning is involved before a campaign against cold can be staged. The required manpower alone is staggering. Mathias's regular crew numbers only about a hundred men, but he needs two hundred men to light and refill his oil heaters. The decision to call out such expensive platoons is made on the basis of teletyped bulletins from the Frost Warning Service, and the service itself is, in the main, a small, spare, superbly professorial man, with a leaky pipe and a tousled mustache, whose name is Warren O. Johnson. His experience with freezing air goes back to his youth; he was born and raised in Benson, Minnesota. He was sent to Florida in 1935 by the United States Weather Bureau after a disastrous forecast, originating in Washington, had predicted warm temperatures for Florida on the night of December 12, 1934, the date of one of the most severe freezes of the twentieth century. The microclimate of Florida, in weathermen's terms, obviously needed its own microforecast, and Johnson has been so successful at it that whenever he gets depressed by other factors in his life he goes out into the groves to talk to people like Mathias, and he says that their praise heals his spirits. Johnson predicts temperatures for dozens of points all over the Ridge and the rest of the state, and his thoughts are superimposed on television screens almost as quickly as they develop. He is seldom more than one degree off in his predictions. He has predicted fifty degrees for one spot and thirty-two degrees for another spot less than a mile distant, hitting it on the nose both ways. When he says that there is going to be a "hard freeze"—twenty-six degrees or lower for four hours or more—growers who have heating equipment get going. He tells them how low the temperature will go and for how long, how fast it will drop, when it will hit twenty-eight degrees, what the differences are likely to be in their area between high and low ground, how calm the night will be, how much frost deposit will come with the cold, what changes in the weather to expect during the night, what temperatures to plan for the next night, and how long the over-all snap will last.

Because of the high cost of heating equipment, Mathias told me, many growers trust to luck and location to get them past a freeze. An unprotected grove that is next to and due south of a heated grove will benefit from the warm air produced at the neighbor's expense, and owners of unprotected groves have actually complained that their northerly neighbors were putting out inadequate amounts of heat.

In spring, as the weather warms and dries out, the overhead sprinkler systems and the portable irrigation pipes on the ground come into their period of more conventional use. After the blossoms go, each tree has many thousands of minute oranges on it—several times the number it could ever conceivably bear to fruition. Taking a twig in his hand, Mathias snapped it off and handed it to me. "Count the fruit right there on that one little twig," he said. The new, deep-green oranges were each about the size of the head of a hatpin. I counted eighteen. "Well, hell, you can't have eighteen oranges on one twig," he said. "It's got to drop some of them. But you want to keep as many as you can. Good moisture sets more fruit. If the weather is dry, you irrigate in an effort to head off a lot of this drop and set a bigger crop."

He picked a mature orange from another tree, opened his knife, and peeled away the orange part of the skin —from the stem end—until the fruit had a kind of snow-capped appearance. He cut a conical hole in the top and, with a circular motion of the knife blade, severed the membranes dividing the segments inside. Juice welled up. "Have some orange juice," he said. The juice came out of the orange almost as if from a glass. "You have to take the orange part of the skin off because otherwise the oil in it burns your lips," he explained, whittling one for himself. I asked him why he opened the hole in the stem and not the blossom end. "If you're going to suck the juice this way, you've got to," he said, "or a lot of times you'll blow that little button and it'll leak on you."

Mathias drove to a block of grapefruit trees. "Let me show you something about grapefruit," he said, getting out. He picked one from one tree and one from another, both wide and flat and, to my eye, almost identical. With two strokes of his knife, he cut off the top and the bottom of one grapefruit, and with several more cuts he chopped off sections of the peel until he had only the edible part in his hand. Setting the blade along one membrane, he cut down to the center, turned the blade, and cut up and out along the next membrane until a section came free. It tasted fairly good, but bland. He cut the other grapefruit in the same way, shook a few seeds away, and offered it to me. It was considerably better than the first one, with an agreeable smartness and character in its flavor. "That was a Duncan," he said. "The other was a Marsh Seedless. There's no comparison, really. But housewives keep demanding the Marsh Seedless because they don't want to worry about seeds."

The canopies of grapefruit trees are higher than the canopies of orange trees, and, looking around, I could see for some distance through the pattern of trunks into what seemed to me, as in the orange groves, to be an unnatural and all but unending silence. I asked him what creatures live in the groves.

"There aren't many," he said. "Rabbits come in and

out. They gnaw the bark of the trees. Wildcats hunt the rabbits. People with dogs hunt the wildcats. Deer eat the leaves. Cows sometimes come in and eat the fruit. There are mice, of course, and rats, and what you call land turtles (we call them gophers), and what you call gophers (we call them salamanders). There are chicken snakes, gopher snakes, and rattlesnakes. They don't live in the groves; they just come in to eat the mice. We used to have a fellow in the crew who was called Snake Man. His real name was Walter Henderson, but no one knew it. He was a tree planter, a sprayer, a grove man. They called him Snake Man because he always caught snakes and scared the others. He couldn't stay away from snakes. One day, Snake Man picked up a little old ground rattler and waved it at the boys. It bit him in the thumb, and the foreman told him he had to go to the hospital. Snake Man said it was nothing and he wasn't going. They finally got him there, and his arm swelled up three times its size, and they thought he was going to die. He was in the hospital a week, and a few days after that he came back to work. When the other boys saw him, one of them called to him, 'Hey, Snake Man, how are you?' Snake Man stopped, and he looked at the whole bunch of them, and he said, 'My name is Walt.' "

I would like to have met Snake Man, but that proved to be impossible. He disappeared a few years ago, and he is thought to have been murdered. I had to settle for

Bird Man, an amiable but untalkative, hardworking orange picker I met one day, whose wife and three-year-old son were in the grove with him. The boy played in the sand beside a galvanized tub, which his mother was filling with oranges to supplement the family income. Orange pickers are an admirable group. They are, in a sense, individual contractors. They work at their own speed and they stay at it according to their own will and endurance. The work they do is so hard that only people of considerable toughness of body and spirit last very long in the groves. Weak, slow, or lazy people make almost no money picking oranges, because their pay is measured by the amount of work they do. Under the Secretary of Labor's plan to reduce unemployment by taking unemployed people to places where work exists, busloads of workers from other states have been sent to Florida during the orange harvests, but nearly all of these people have quit at once. One of them wrote to the Florida State Employment Service, saying, "In one half day, I have learned that I am not a citrus picker. If you know of any seasonal farm work in the state that doesn't require the combined agility of a monkey and the stamina of a horse, I would like to hear of it." The working day for pickers begins when the dew dries, because the skins of oranges grow taut when covered with water and the fruit is easily damaged. Charles Baty, the foreman of the crew in which Bird Man was working, told me that Bird Man is an average picker and that, with his wife's help, he gets about eighty boxes of oranges a day. At twenty-five cents a box, that makes twenty dollars. Sometimes, when a picker comes to a tree, a good many oranges are already on the ground. Many seem perfectly healthy. Fallen fruit is not picked, however, since it can raise the bacteria count beyond the levels permitted by government inspectors in packinghouses and concentrate plants. Some growers go through their groves with disk harrows just before picking time in order to chop up the fallen fruit and relieve pickers of the temptation to stuff their bags with it. A picker has to pick each tree clean. An isolated orange that has been missed is called a shiner, because of the way it will shine like a light bulb alone in the tree. Baty goes around with a shiner pole in his hand. It has a hook on one end. If only one or two oranges are left in a tree, he will flick them out with the pole. If he finds a half-dozen shiners, he will send the picker back to the tree. "If I don't send a man back for six, the next time he'll leave twelve," he explained. At that moment, he saw a tree with an array of shiners glistening within its canopy. "Bird Man," he called out, "you got about eight or ten up here." Bird Man got out of the tree he was in and picked the shiners, looking guilty.

"You should talk to Doyle Waid," Baty said to me. "He's through for today because the strap of his picking bag cut his shoulder. I've never seen anyone better than Doyle. He's quit early and he's already picked a hundred boxes anyway."

Waid, a short and somewhat frail man with sandy

hair and shallow, light-blue eyes, seemed to weigh about a hundred and thirty-five pounds. The back of his neck was heavily creased from his years under the sun. He lived, at the time, in a trailer in Davenport, Florida, with his wife, his four sons, and his daughter. He said he picked oranges six days a week and on Sunday as well when there was work. Until he was eighteen, he lived in Cleveland, Mississippi; then he came to Florida, started picking, liked the money, and stayed, leaving the state only in the summer to work apples, cherries, and pears, near Hartford, Michigan. He is white. About three-quarters of Florida's orange pickers are Negroes. Standing there drinking a cup of water that particular afternoon, he looked like someone who had earned his money. He looked worn out. He had no shirt on, and his body was glistening in the heavy humidity and heat, which was in the nineties. There was a red welt across his shoulder and down his chest, where the strap of his picking bag had rubbed through his skin—enough, in places, to bring out some blood. Gnats were swarming all over the welt, but he paid no attention to them. It was five minutes past three, and he had picked twenty thousand oranges that day. He would get twenty-five dollars, and he seemed to be feeling pretty good about it. Picking a hundred boxes a day is impossible for most people, but for him it had become a minimum standard. After all, he was good at it, and he was in his prime as a picker. At the time, he was twenty-nine years old.

"I have made as high as forty-one dollars in a day," Waid told me.

"When did you do that?"

"Back in tangerines."

Each tangerine has to be removed from the tree with a pair of clippers, or a plug will pull out of the skin. Tangerines pay three times as much as oranges, which are snapped from the tree with a fast turn of the wrist that is similar to the motion used by a baseball pitcher throwing a curve. In California, oranges are clipped like tangerines. Florida gave up clipping oranges during the Second World War. Pickers try to fill their picking bags each time they make one descent on their ladders. Certain historic orange trees have borne as many as twelve thousand oranges at one time, but an average yield is closer to fifteen hundred. Waid pointed out a nicely *per tree* structured tree with well-spread limbs. "That tree is good picking," he said. "It's got big fruit. It's easy to ladder. You can set your ladder in there and get your inside fruit good. That's a bad tree next to it, with the branches close together. It's a hard set. A good average tree, it oughtn't to take over seven sets to get it."

The main direction of research in harvesting seems to be toward getting rid of human pickers. With the rise of concentrate, orange-tree plantings are multiplying, but the picking force is diminishing. Florida used to depend on the supplementary help of offshore labor—pickers mainly from the British West Indies—but the

Department of Labor is making it increasingly difficult for these workers to enter the United States. So, with a sense of considerable urgency, citrus men are hoping for the development of a mechanical harvesting machine. The citrus business is looking for an Eli Whitney, and many candidates are applying themselves to the pursuit of the fortune that would settle upon the inventor of what might be called the orange gin. One is Fred D. Lasswell, Jr., of Tampa, who draws the comic strip "Barney Google and Snuffy Smith." Lasswell's entry is a big set of whirling flexible fins, which are pushed up against a tree so they can snap the oranges free from the twigs that hold them. International Harvester once tried to develop Lasswell's fins, but apparently without great success. William Adams, the nurseryman I visited, has something in a back room that looks like a giant version of the conical spring that fits behind the batteries in an ordinary flashlight; spiraling through the branches of a citrus tree, it is supposed to disengage the fruit with gentle force. Another inventor has tried sucking the fruit off the tree by using a kind of vacuum cleaner, more or less turning the picker into a janitor.

The most active of the new Whitneys—and at the same time the central coördinator of nearly every other inventor's efforts in the state—is Glenn Coppock, an agricultural engineer at the Experiment Station. Coppock, a public servant whose salary comes from the Florida Citrus Commission, tries out all sorts of new ideas, his own and those of others. One of his inventions is an assemblage of outsize, soft-rubber drill bits that screw themselves into the canopy of a tree and, on the way in, twist off the oranges. He has also tried combing the trees with huge combs. It takes a man about an hour to pick an average tree clean. Coppock wants to get all the fruit off the tree in a couple of minutes. Recently, he has been rolling into the groves with something he calls the Inertia Tree Shaker. This is an enormous steel arm, eighteen feet long, with a padded claw on one end, that reaches into a tree, grabs hold of a branch, and thunders back and forth while the oranges drop into a nylon apron. Tree shakers are already in use in the harvesting of nuts and of certain fruits that can stand a little damage—such as cherries for canning and plums that are destined to be prunes. Oranges headed for concentrate plants could be harvested in this way, too. Coppock is also working on what he calls the Oscillating Air Blast Machine. This brute is simple enough. It is a bank of huge fans that goes up to a tree and blows at it with winds of hurricane force. The air emerges through louvers, which move so that the wind is delivered in gusts that swing oranges like pendulums and tear them free. The Oscillating Air Blast Machine can clean a tree in two minutes. However, the defoliation of the tree may have been so considerable that it will appear to be in the final throes of spreading decline. One major problem for both the shakers and the air-blast machines is that Valencia trees, half of the total number in the state, carry ripe fruit and new fruit at the same time, and therefore can be neither shaken nor blasted at harvest time without destroying the following year's crop. Coppock feels that the ultimate device toward which all these intermediate efforts are leading will be a vast behemoth on wheels, built in the shape of an A-frame, which will straddle a row of A-shaped trees, spraying when necessary, pruning when necessary, and somehow harvesting the fruit—probably by some obvious and uncomplicated system, which has yet to form in a human brain.

I asked Doyle Waid if he had ever seen one of Coppock's mechanical pickers.

"I seen 'em work," he said. "They don't work."

THREE

CITRUS SINENSIS

A TRACTOR driver gave me a ride out to my car when I left the picking crew one day, and I've forgotten what I said to him when I jumped up behind him, but whatever it was, he turned around with a look of recognition and said, "You come from apple country." In one sentence, he had defined the dimensions of his own world, the utterly parochial nature of it, its disciplined singleness, its weather, and its cycles of fruition. The appeal of that world and, to an even greater extent, the relief of it had increased in my mind with each day in the groves, among other reasons simply because gas stations, Burger Queens, and shopping centers so dominate the towns of central Florida that the over-all effect on a springtime visitor can be that he is in Trenton during an August heat wave. The groves, in absolute contrast, are both beautiful and quiet, at moments eerie. I retreated into them as often as I could. To someone who is alone in the groves, they can seem to be a vacant city, miles wide and miles long.

In late afternoons and early evenings, I sometimes stayed there to read, sitting on the sand against the trunk of a tree, thumbing through books that had been recommended to me by pomologists at the University of Florida's Citrus Experiment Station. They dealt with the history and botany of citrus, the physiology of the orange, citrus growing, citrus products, and citrus industries. The most absorbing was an encyclopedic treatise called *Hesperides: A History of the Culture and Use of Citrus Fruits*, written in the nineteen-thirties, in what was then Palestine, by a scholar named Samuel Tolkowsky and titled as a gesture of respect for the work of the Sienese priest, Giovanni Battista Ferrari, whose own *Hesperides* had appeared nearly three hundred years earlier. Tolkowsky's work included a kind of compendium of everything of importance—including its namesake—that had been published on the subject in earlier centuries. Thomas Cardinal Wolsey, I found, had also warded off the unpleasant aspects of society by retreating into oranges. Going from place to place in sixteenth-century London, he liked to carry in his hand an orange that had been capped and hollowed out. Inside it, he would put a bit of sponge, saturated with vinegar. With his nose resolutely pressed into the orange, he was insulated by the aroma of the peel and the vinegar against the noxious airs of London and his fellow men.

The evolution of citrus probably began in the Malay

Archipelago at least twenty million years ago, when the islands of the South Pacific were still part of a body of land that included Asia and Australia. A bitter ancestral plant apparently made its way to what is now the Asian mainland, and from it developed the modern fruit. The evidence that this event occurred in the area of southern China is overwhelming, beginning with the fact that more citrus varieties and more citrus parasites can be found there than anywhere else. Spreading out to the rest of the world, Chinese citrus jumped the East China Sea and reached Japan by way of Formosa and intervening island groups. It moved eastward into the South Pacific, and the frequency of citrus in the islands today diminishes with distance from the mainland. In the junks of merchant seamen, seeds and trees were carried south to the shores of the Java Sea and into the Strait of Malacca, which was a kind of departure point for sometimes unexpected migrations to India and Africa on the strong westward currents of the Indian Ocean.

Among the first citrus varieties to make this journey —and then to go on into the Mediterranean basin—was the citron, which acquired its name because of an early confusion with another tree. According to Tolkowsky, this confusion came about because the large, rough-skinned citron resembled the greenish-yellow cone of the cedars of Lebanon, and since citron trees and orange trees are nearly identical in shape and foliage, the confusion inevitably expanded. The Greeks called the citron a *kedromelon*, or "cedar apple." The Romans turned this into *malum citreum*, and applied the term—often shortened merely to *citreum*—to all of the varied fruits of citrus trees. The second-century writer Apuleius, for one, objected. He had been born in Africa and knew a cedar cone from an orange. In the eighteenth century, the Swedish botanist Carolus Linnaeus nonetheless made the name "citrus" official for the genus. So lemons, limes, citrons, oranges, grapefruit, and tangerines are now grouped under a name that means cedar.

The word "orange" evolved from Sanskrit. The Chinese word for orange, in ancient as well as modern Chinese, is *jyu*, but it did not migrate with the fruit. India was the first major stop in the westward travels of citrus, and the first mention of oranges in Sanskrit literature is found in a medical book called the *Charaka-Samhita*, which was compiled approximately two thousand years ago. The Hindus called an orange a *naranga*, the first syllable of which, according to Tolkowsky, was a prefix meaning fragrance. This became the Persian *naranj*, a word the Muslims carried through the Mediterranean. In Byzantium, an orange was a *nerantzion*. This, in Neo-Latin, became variously styled as *arangium*, *arantium*, and *aurantium*—eventually producing *naranja* in Spain, *laranja* in Portugal, *arancia* in Italy, and *orange* in France.

Meanwhile, the Roman city of Arausio, in the South of France, had become, in the Provençal language, Aurenja—a name almost identical in sound and spelling to *auranja*, the Provençal word for orange. Gradually, the names of the city and the fruit evolved in the Provençal tongue to Orenge, and then to Orange. In the early six-teenth century, Philibert of Orange, prince of the city, was awarded a good part of the Netherlands for his political and military services to the Holy Roman Em-peror, Charles V. The Prince had no immediate heir, and his possessions and titles eventually passed to a German nephew. This was William of Nassau, Prince of Orange, who founded the Dutch Republic and the House of Or-ange. In honor of William's descendants, Dutch explor-ers named the Orange River, in South Africa, and Cape Orange, in northern Brazil. Fort Orange was the name of a Dutch settlement that eventually developed into Albany, New York. After a Protestant prince of the House of Orange had served as King William III of England, a movement known as Orangeism was founded by Irish Protestants, who established the Orange Society, and even called their part of Ireland "The Orange." Commemorating their cause on the landscape of the New World, emigrant Orangemen gave the name "Or-ange" to towns, cities, and bodies of water, from Lake Orange, Maine, to Orangeburg, South Carolina. Orange-men changed the name of Newark Mountains, New Jersey, to Orange Dale, which eventually became simply Orange, New Jersey, with its satellite towns of West Orange, South Orange, and East Orange—all as the re-sult of a similarity of sound between the name of a trans-alpine Roman city and the name of a citrus fruit.

Nominal confusion also resulted from a tendency among Romans and Greeks to call any kind of fruit an apple. When the Romans discovered the pomegranate in Punic Mauretania—now Morocco and Algeria—they called it the *malum punicum*. When they came upon the peach, in Persia, they called it the *malum persicum*. Centuries earlier, in Media and in Persia, botanists traveling with the conquering armies of Alexander the Great had found the citron and had named it, variously, the Median apple and the Persian apple. Working later with material left by Alexander's scientists in the archives of Babylon, Theophrastus, the greatest of Greek botanists, also described citrons as Persian and Median apples, and his work disseminated the terms throughout the ancient world. It was a "golden apple" that Paris gave to Aphrodite, thus opening his way to the heart of Helen. In Antiphanes' *The Boeotian Girl*, written in the fourth century B.C., a young man presents a citron to his mistress, and she says,

> "I thought it came from the Hesperides,
> For there they say the golden apples grow."

Other Greeks, it appears, thought that the golden apples were quinces. Tolkowsky points out that a frieze in the Temple of Zeus at Olympia shows Herakles holding a handful of quinces. In Rome, however, universal agreement seems to have been reached that the golden apples were citrus.

According to Father Ferrari, the Romans thought that citrons, oranges, lemons, and other citrus fruits came to Italy in the arms of the Hesperides—the daughters of

Hesperis and Atlas—who crossed the Mediterranean from Africa in a giant shell. Oranges actually reached the Italian peninsula from India. In the first and second centuries A.D., it was only a seventy-day trip across the Indian Ocean from the Malabar Coast to the western shore of the Red Sea, twelve more days from Berenice by camel to the Nile, and another twelve down the river to waiting galleys at Alexandria. (Orange groves were established at Berenice and elsewhere on this route, which eventually branched into the Levant.) Toward the end of the Roman Empire, oranges were flourishing on the Italian peninsula. After the fall of Rome, oranges played a part in the great Lombard invasion. A Byzantine governor of Rome, enraged at being summarily called back to Byzantium, sent an embassy with a selected display of Italian oranges to Alboin, King of the Lombards, inviting him to overrun Italy, which Alboin did.

In the sixth and seventh centuries, the forces of Islam conquered a wide corridor across the world from India to Spain, and orange, tangerine, and lemon trees today mark the track of the Muslim armies. After Moorish capitals had been established in Andalusia, desert artisans and architects, delirious in the presence of water, filled and surrounded their buildings with pools, cascades, and fountains, planting a small grove of oranges in the Great Mosque of Cordova and oranges and lemons in the interior courts of the Alhambra in Granada. One curious footnote to the rise of Islam developed in Italy in the eleventh century. A group of Norman pilgrims, on their way home from the Holy Land, came upon a band of warrior Muslims who were about to destroy the person and possessions of a Christian prince of Salerno. The Normans saved the prince and drove the Muslims away. Fearful of further attacks, the prince, like the Byzantine governor of Rome nearly five hundred years before him, sent an embassy with the pilgrims to the Duke of Normandy, accompanied by a mountainous gift of beautiful oranges, frankly tempting the Duke to conquer southern Italy—which he did, taking Sicily, too. The Norman conquest of Sicily turned into something of a scandal. Norman minds dissolved in the vapors of Muslim culture. Austere knights of Honfleur and Bayeux suddenly appeared in the streets of Palermo wearing flowing desert robes, and attracted to themselves harems of staggering diversity, while the Church raged. Norman pashas built their own alhambras. The Normans went Muslim with such remarkable style that even Muslim poets were soon praising the new Norman Xanadus. Of one such place, which included nine brooks and a small lake with an island covered with lemon and orange trees, the poet Abd ur-Rahman Ibn Mohammed Ibn Omar wrote:

> The oranges of the Island are like blazing fire
> Amongst the emerald boughs
> And the lemons are like the paleness of a lover
> Who has spent the night crying . . .

It is only in comparatively recent centuries that oranges, in Western countries, have actually been eaten as a food. Their earliest popularity in Europe seems to have been based on the ornamental appearance of the trees and the inspiring aroma of the peel and the blossoms. At the table, they were used as a seasoning for meat and fish and seldom consumed in any other way. Before 1500, European orange growers mainly grew Bitter Oranges, because they were more aromatic, better as seasoning, and hence more valuable. Dinner guests could measure their importance in the regard of their hosts by the number of oranges that came to the table. One fourteenth-century cookbook, describing a dinner given by an abbott of Langy for his superior, the Bishop of Paris, indicates how impressive a meal it was by noting that the roast fish was seasoned with powdered sugar and Sour Oranges. In 1529, the Archbishop of Milan gave a sixteen-course dinner that included caviar and oranges fried with sugar and cinnamon, brill and sardines with slices of orange and lemon, one thousand oysters with pepper and oranges, lobster salad with citrons, sturgeon in aspic covered with orange juice, fried sparrows with oranges, individual salads containing citrons into which the coat of arms of the diner had been carved, orange fritters, a soufflé full of raisins and pine nuts and covered with sugar and orange juice, five hundred fried oysters with lemon slices, and candied peels of citrons and oranges.

At about that time, Portuguese ships returned home from India with sweet orange trees, and a new type spread through Europe. It became known as the Portugal Orange, and it quickly replaced the Bitter Orange in popularity throughout the continent. The word "Portugal" became synonymous with good sweet oranges in numerous countries, and, in fact, sweet oranges are still called Portugals in Greece, Albania, Rumania, parts of the Middle East, and some parts of Italy.

In most of Western Europe, the favor held by the Portugal Orange was less enduring. Within a century after the first trees had come from India, Portuguese missionary monks sent word back from China that Chinese oranges were sweeter than sugar itself. One Portuguese Jesuit wrote that "the oranges of Canton might well be muscat grapes disguised." In 1635, a Chinese orange tree reached Lisbon, and before long the China Orange—a term still used in many countries to denote a fine sweet orange—was in demand all over Europe. The botanical name of the modern sweet orange, in fact, is *Citrus sinensis*.

FOUR

ORANGERIES

Seventeenth-century Frenchmen used to pummel oranges, then heat them over glowing coals in order to extract as much juice as possible. They particularly liked to pour the juice over their roast chestnuts. A drop in the price of West Indian sugar inspired the invention, in Paris in 1630, of lemonade. Orangeade followed, but was not as cheap or as popular. Owners of the sidewalk cafés of Paris became known as *limonadiers*. They sold coffee and other drinks, as they do today, but for about two centuries they were known both popularly and officially as *limonadiers*.

In fifteenth-century Breslau, there was an annual orange shoot—the *Pomeranzenschiessen*—during which marksmen spent happy hours shooting oranges off of one another's heads. In Switzerland, the legend of William Tell was recorded in the same century, and since there seems never to have been a William Tell, it is possible that the Swiss borrowed their idea from the Breslovians, and that the fruit on the head of Tell's trusting son might well have been an orange.

Archers of the Chou Dynasty made their bows of the wood of orange trees. Very tough, with a fine and straight grain, no burl, and a tinge of orange color, orangewood has been used for inlays (as well as for whole pieces of furniture) by craftsmen in all centuries. A set of orangewood points was the forerunner of the modern toothbrush. And for more than three thousand years orangewood also continued to be a favored choice for the bows of archers, but it has been largely replaced —in the United States, at least—by fiber glass.

Oranges and orange blossoms have long been symbols of love. Boccaccio's *Decameron*, written in the fourteenth century, is redolent with the scent of oranges and orange blossoms, with lovers who wash in orange-flower water, a courtesan who sprinkles her sheets with orange perfume, and the mournful Isabella, who cuts off the head of her dead lover, buries it in an ample pot, plants sweet basil above it, and irrigates the herbs exclusively with rosewater, orange-flower water, and tears. In the fifteenth century, the Countess Mathilda of Württemberg received from her impassioned admirer, Dr. Heinrich Steinbowel, a declaration of love in the form of a gift of two dozen oranges. Before long, titled German girls were throwing oranges down from their balconies in the way that girls in Italy or Spain were dropping handkerchiefs. After Francis I dramatically saved Marseilles from a Spanish siege, a great feast was held for him at the city's harborside, and Marseillaise ladies, in token of their love and gratitude, pelted him with oranges. Even Nostradamus was sufficiently impressed with the sensual power of oranges to publish, in 1556, a book on how to prepare various cosmetics from oranges and orange blossoms. Limes were also used cosmetically, by ladies of the French court in the seventeenth century, who kept them on their persons and bit into them from time to time in order to redden their lips. In the nineteenth century, orange blossoms were regularly shipped to Paris in salted barrels from Provence, for no French bride wanted to be married without wearing or holding them.

Paradoxically, many societies have believed that the worst thing that could happen to an orange tree was the touch of a woman. If a woman were even to go near one, some thought, the foliage would wilt and fall away, the fruit would drop, and the tree would die. A Spanish Moor of the twelfth century, whose name was Abu Zakariya Yahya Ibn el-Awwam, wrote a basic text called *The Book of Agriculture*, which contained material on citriculture that was remarkably accurate and complete, until he brought up the matter of women. "Women should not be allowed to come near citrus trees," he wrote, "unless they are in a state of absolute purity and unimpaired health." According to the same writer, however, the woman stood to gain much from the very tree she was capable of destroying. "If a woman eats an orange," he added, "it will banish all evil thoughts from her mind." Superstition about oranges was remarkably persistent in Germany. As late as 1671, Italian orange salesmen found that Frankfurt was a poor territory. The Frankfurt City Council proclaimed that the salesmen were going around spreading "poisonous yellow ointments" on the doors of houses. If a person passed through or even near these doors within five hours, the council informed the populace, that person would die. German feelings about women and oranges were even deeper. In the early eighteenth century, when nearly all German princes were growing oranges in their palaces, Johannes Volckamer, of Nuremberg, in his *Neurenbergische Hesperiden*, described how women could cause whole trees to die. "Many will deride this as something foolish," said Volckamer, "and I myself should not have believed in it had it not caused the undoing of some of my most valuable trees. Once, in winter, I noticed a woman of my gardener's household seated upon a beautiful orange tree in full bloom. The next day, the tree started drying up from the top downwards, and so rapid was the progress of the disease that in the course of a few days it had infected every single branch, causing all the leaves to wilt and die."

Later writers have guessed that Volckamer was ignorant of the effects of frost. My own belief is that science erases what was previously true. The earth *was* the center of the universe until Copernicus rearranged it. Life *did* begin in Eden before Darwin restyled it. In the early eighteenth century in Nuremberg, a woman *did* sit in the branches of an orange tree and kill it to the ground.

In ancient civilizations, the juice and peel of oranges and other citrus fruits were prescribed as antidotes for innumerable poisons. The conviction that the fruit was somehow useful in the cause of human health continued to develop during the feudal ages, and in the middle of the twelfth century the remedial role of the orange was given professional ratification in a Moslem medical textbook. The Tunisian Doctor Abu Abdullah Mohammed ben Mohammed el-Huseiny el-Ali Billah, in his *Treatise of the Simple Remedies,* said that powdered orange peel stirred into hot water would stop an attack of colic at once. To get rid of a tapeworm, he wrote, one had only to prepare an emulsion of powdered orange peel and olive oil. To combat "poison of a cold nature," he said, drink wine to which the powdered root hairs of orange trees have been added. During the Italian Renaissance, physicians pondering the threat of plagues decided that the orange was the ultimate preventive. The Florentine philosopher Marsilio Ficino, leader of Cosimo de' Medici's Neo-Platonic Academy, persuaded people throughout the state to carry oranges in their hands to keep the plague at bay. In 1544, Petrus Andreas Matthiolus decided that even orange-flower water, long in use as a perfume, was an excellent specific against the fevers of the plague, a belief which may at least have kept him and his followers in a healthy frame of mind until 1571, when he died of the plague.

With the age of sail, the antiscorbutic value of citrus fruits had been discovered. This came about when sailors noticed that if they ate limes, lemons, and oranges in the course of long voyages, the livid splotches went away around the roots of their hair, their muscles stopped aching, their skins regained color, their once appalling breath grew sweeter, the swelling in their legs went down, bruises vanished, new strength replaced a feeling of deadening fatigue, their gums stopped bleeding, their teeth stopped dropping out, their bones stopped breaking, their hair began to grow again, and their spirits rose. Portuguese sailors in the sixteenth century carried citrus trees to St. Helena and covered the island with them in order to establish a kind of scurvy-prevention station for early and late stops in their prodigious voyages on the route that Vasco da Gama had opened to the Orient. Portuguese, Spanish, and Arab crews planted similar clinics on the African west coast, in the Azores, and in the Madeira island group. The Dutch established a citrus plantation in South Africa in 1654. After the British Admiralty issued orders for regular rations of lime juice on all of His Majesty's ships, British sailors became known as limeys.

The healthful properties of citrus fruit have since been found to cover a far wider range of ills than nautical scurvy. Oranges, according to the kind of reading matter that comes in a doctor's mail, are salubrious for people with peptic ulcer, obesity, burns, liver trouble, arthritic pain, pernicious anemia, megaloblastic anemia, iron-deficiency anemia, gastric malignancy, diarrhea, rheumatic heart disease, rheumatic fever, acne, osteomalacia, hypoglycemia, multiple sclerosis, strokes, heart attacks, periodontitis, gingivitis, and fatigue. With the help of oranges, say the pamphlets, bones grow strong, sound teeth and gums are built, wounds heal rapidly, miscarriages are prevented, the damage and duration of bruises are limited, temporarily depleted blood-sugar reserves are swiftly restored, and the mind's ability to withstand stress is increased. Football players who are full of orange juice get up a little faster after a play. They get put out of action less often, and if they do happen to get injured, they recover more quickly. I have asked several doctors about all of these claims, and they agreed that, in one way or another, the claims are essentially valid. They also said, however, that the medical benefits of orange juice are generally overemphasized today. Scurvy, when it occurs, is now easy to put down. And in all instances it is only Vitamin C, after all, that is doing the work, and Vitamin C can be readily absorbed from capsules.

The modern use of lemon and orange peels in alcoholic drinks has ample precedents in many places and centuries, but the custom perhaps reached its highest point in Holland three hundred years ago. The drinking Dutch, in the seventeenth century, would peel a helical ribbon of skin from an orange, continuing round and round until the knife reached the fruit's equator. Then, with the ribbon still attached, they would place the entire orange, like a huge Martini olive, in the bottom of a wineglass so large that it might have been described as an elegant bucket. Dutch fondness for the combination of oranges and wine eventually led to the invention of bitters—or at least to the commercialization of bitters, for even the ancient Chinese had known the special excellence of Bitter Oranges with wine. Dutch bitters were a concentrated essence usually made by marinating dried Bitter Orange peels in gin. The still-lifes of Dutch and Flemish masters often show oranges beside bottles of wine.

Because of a combination of new artistic techniques and some apparently reasonable, but mistaken, assumptions about the history of citrus, oranges appeared frequently in paintings by any number of the great masters of the Italian Renaissance. In making the break from Byzantine scholasticism to the new humanism of the Renaissance, artists began setting their religious figures against naturalistic backgrounds. Not having seen the Holy Land, they glibly set their Annunciations and Resurrections in Italian villas and on Italian hills. Crusaders, among others, had long since reported that orange trees flourished in Palestine, so, as a kind of hallmark of authenticity, the painters slipped orange trees into masterpiece after masterpiece, remaining ignorant to their deaths that in the time of Christ there were no orange trees in or near the Holy Land.

In his "Maestà," the Sienese painter Duccio di Buoninsegna showed Jesus entering Jerusalem through the streets of Siena, past orange trees in full fruit. Fra Angelico painted Jesus resting under an orange tree. It was almost unthinkable for a great master to do a "Flight into Egypt" without lining the route with orange trees. A "Last Supper" was incomplete without oranges on the table, although there is no mention of oranges in the Bible. Titian's "Last Supper," which hangs in the Escorial, shows oranges with fish. A Domenico Ghirlandaio "Last Supper" goes further: a mature orange grove is depicted in murals behind the Disciples. The deterioration of Leonardo's "Last Supper" has been too extensive for any oranges in it to be identified, but in all likelihood, according to Tolkowsky, they were there. Most painters thought of the Annunciation as occurring indoors, and Paolo Veronese, for one, moved orange trees indoors to authenticate the scene, setting the plants in trapezoidal pots, of the type in which orange trees were grown in his time in northern Italy. Fra Angelico also used orange trees to give a sense of the Holy Land to his "Descent from the Cross," which was otherwise set against the walls of Florence, and, like many of his contemporaries, when he painted the Garden of Eden he gave it the appearance of a citrus grove. Benozzo Gozzoli's frescoes in a family chapel of the Medici show Melchior, Baltha-sar, and Gaspar looking less like three wise kings from the East than three well-fed Medici, descending a hill that is identifiable as one near Fiesole, dressed as an Italian hunting party, and passing through stands of or-ange trees bright with fruit. Actually, Gozzoli's models for the magi were Lorenzo de' Medici; Joseph, Patriarch of Constantinople; and John Palaeologus, Emperor of the East.

The orange tree was more than a misplaced landmark. It was also a symbol of the Virgin, erroneously derived from an earlier association that medieval theologians had established between Mary and the tall cedars of Lebanon. Thus, countless paintings of the Madonna or of the Ma-donna and Child were garlanded with orange blossoms, decorated with oranges, or placed in a setting of orange trees. Mantegna, Verrochio, Ghirlandaio, Correggio, and Fra Angelico all complemented their Madonnas with or-anges. Sandro Botticelli, in his "Madonna with Child and Angels," set his scene under a tentlike canopy thickly overhung with the branches of orange trees full of oranges.

In the Neo-Latin of the Renaissance, oranges were sometimes called *medici*—an etymological development that had begun with the Greek word for citron, or Median apple. It is no wonder, therefore, that in art and interior decoration the Medici themselves went in heavily for oranges. In Florence, oranges are painted all over the ceilings of the Medici's Pitti Palace. The Grand Duke Francesco de' Medici was one of the world's earliest collectors of citrus trees, and in Tolkowsky's view the five red spheres on the Medici coat of arms were almost certainly meant to represent oranges. When Botticelli painted his "Primavera," under a commission from the family, he shamelessly included Giuliano de' Medici as the god Mercury, picking oranges. Botticelli also painted his "Birth of Venus" for the Medici. It was Venus, and not the Hesperides, according to a legend current at the time, who had brought oranges to Italy. Botticelli's model was Simonetta dei Cattanei, wife of Marco Vespucci and Giuliano de' Medici's Platonic love. Simonetta came from Porto Venere, where Venus was alleged to have landed with the original oranges, so Botticelli painted her in the celebrated scallop shell bobbing on the gentle swells off Porto Venere, and lined the coast behind her with orange trees. Giuliano's son, Giulio de' Medici, who became Pope Clement VII, commissioned Raphael to design a villa for him with a great double stairway leading to a sunken garden full of orange trees.

All this was bound to engage the envy of royalty in the north, and at the end of the fifteenth century, in an expedition often said to mark the dividing point between medieval and modern history, Charles VIII of France went to Italy intending to subdue the peninsula by force of arms. Instead, he fell in love with Italian art, architecture, and oranges. When he returned to France, every other man in his retinue was an Italian gardener, an Italian artist, or an Italian architect. Charles was going to transform the castles and gardens of France.

Oranges had been grown before, to a limited extent, in northern climates. In the first century, the poet Martial had noted that some growers in Italy had protected their trees from frost by sheltering them under mica. Growers of the third century had regularly planted their trees in semi-sheltered places. Charles took these ideas home with him, and, at his château at Amboise, built the first *orangerie*. His wife, Anne of Bretagne, then built her own *orangerie* at her château twenty-five miles away, in Blois.

For the next two hundred years, a French reign was incomplete unless the king had built an *orangerie* larger and more magnificent than the one built by his predecessor. Most were vaulted galleries, open on the southern side. Horticulturally, they were merely the prototypes from which modern greenhouses would evolve, but architecturally they soon became as celebrated as some of the world's better tombs, palaces, castles, and cathedrals. When Henry II married Catherine de' Medici in

1533, her arrival in France added stimulus to the already expanding boom in Italianate palaces built with accommodations for indoor groves of Italian oranges. The *orangerie* that Henry later built at Anet for his mistress Diane de Poitiers has been pointed to as a summary of the genius of French architecture. Queen Marie de' Medici, lonely in Paris after the death of Henry IV, built the Luxembourg palace in imitation of her family's Pitti Palace in Florence, surrounding it with an imitation of the Medicis' Boboli Gardens, filled with potted orange trees. Her antagonist, Cardinal Richelieu, fought back with a lemon house, which he added to his castle at Rueil. The trees in most French *orangeries* were planted in giant boxes and moved in and out of doors with the seasons. Sometimes the boxes had axles and wheels, so that the practical French could roll their orange trees out for an airing on a sunny winter morning, or move them out of the shadows on a cool spring afternoon. The trees were not planted as seeds but were shipped in winter in balls of earth from Italy on an eight-week voyage to ports like Cherbourg and Le Havre. On receiving them, royal gardeners would bathe the roots in milk or honey. No one has ever loved oranges and orange blossoms more than Louis XIV. His *orangerie* at Versailles was built in the shape of a C, twelve hundred feet around, and was the scene of garden parties and masked balls. His gardeners learned how to parch the trees to the point of death and then dramatically revive them into bloom; by doing this in planned cycles they provided the Sun

King with blossoms throughout the year. There were blossoms all over the palace at all times, supplemented by paintings of oranges and by Beauvais tapestries of oranges.

The French enthusiasm was shared to some extent throughout northern Europe, and there were royal *orangeries* from Stockholm to St. Petersburg. Other people took up the practice of orange-growing, too. The coat of arms of the Bagolini family of Belgium showed Faith making orange juice. Peter Paul Rubens painted himself, his wife, and his son relaxing among the orange trees in their garden at Antwerp. But nowhere was envy of French royal citriculture more obvious and competitive than in Germany. The Count Palatine Frederick V converted a two-hundred-and-eighty-foot terrace outside his castle at Heidelberg into an *orangerie* in 1613, putting over it a roof that was supported by soaring arches, some of which were eighty feet high. He put walls of windows in the arches, and he kept stoves going all winter long to convince his trees that they were really in the tropics. Nothing, however, approached in size or pretension an *orangerie* in Dresden called the Zwinger. *Orangeries* had begun as fairly simple structures—a roof supported by columns—to protect orange trees, but the example of Versailles made it clear that they would end as royal amusement parks containing facilities for banquets and balls, held between stands of orange trees to create the impression that the assembled company was dining and dancing in the middle of an orange grove.

The Zwinger was the end. Named for Zwinger Square in Dresden, which it faced, it was four-sided and larger than the *orangerie* at Versailles. In addition to its banqueting space, it had a theater and a nymphaeum with a waterfall. The Zwinger was built in the early eighteenth century by Frederick Augustus I, Elector of Saxony, who intended it to be an indication of the size of the palace that would go with it, but he ran out of money and left only the Zwinger and its oranges behind him when he died.

In Tudor times, oranges grew in the soil of Britain. Sir Francis Carew planted orange trees in Surrey in 1562. He covered them in winter with a board hut, and they lived for a hundred and seventy-eight years before they were killed to the ground by the cold winter of 1740. England was importing enough oranges from Portugal in the Tudor era to keep barons and upward reasonably well supplied. Shakespeare may have disliked oranges, or perhaps they seldom found their way down to his level; at any rate, in his work he only mentions them four times in passing. One is in a curious simile in *Much Ado About Nothing*, when Beatrice says, "The Count is neither sad, nor sick, nor merry, nor well; but civil Count, civil as an orange, and something of that jealous complexion." During Shakespeare's lifetime, or perhaps a little before him, English children began to chant a nursery rhyme that includes the line "Oranges and lemons say the bells of St. Clement's"—and they have not stopped to this day. The Church of St. Clement Danes, on the Strand near the Inns of Court, was established by Danes living in London in the ninth century, and they equipped it with its remarkably articulate bells. On the thirty-first of March each year, modern Danes living in London load up the church with oranges and lemons and invite hundreds of children to come for a service, giving each a lemon and an orange to take home.

On March 9, 1669, Samuel Pepys had his first glass of orange juice, and, afterward, he wrote in his diary, "I drank a glass, of a pint, I believe, at one drought, of the juice of oranges, of whose peel they make confits, and here they drink the juice as wine, with sugar, and it is a very fine drink; but, it being new, I was doubtful whether it might not do me hurt." Titled Englishmen of the seventeenth century had their own *orangeries*. Lord Sunderland had one in Northamptonshire. Lord Arlington had one at Euston. The Duke of Lauderdale maintained one at Ham. Sir William Temple claimed that his trees gave him all the oranges he cared for and better ones than any save "the best sorts of Sevil and Portugal." Henrietta Maria, queen to Charles I and daughter of Marie de' Medici and the French King Henry IV, kept sixty orange trees in tubs at Wimbledon. Also during the Restoration, young women carrying baskets of oranges used to stand near the stage in London theaters, face the audience, and sell oranges at sixpence apiece and themselves for a little more. The girls were known as Orange Girls, and they worked under the administration of women called Orange Molls. Nell Gwyn, a beautiful and illiterate Orange Girl, became a minor actress and the mistress of King Charles II. "Anybody may know she has been an orange wench by her swearing," said the Duchess of Portsmouth. Nell Gwyn died when she was thirty-seven, but she lived to see her son made Duke of St. Albans.

FIVE

INDIAN RIVER

In the seventeen-seventies Londoners developed a craving for Jesse Fish oranges. These had thin skins and were difficult to peel, but the English found them incredibly juicy and sweet, and Jesse Fish oranges were preferred before all others in the making of shrub, a drink that called for alcoholic spirits, sugar, and the juice of an acid fruit—an ancestral whiskey sour. More than sixty-five thousand Jesse Fish oranges and two casks of juice reached London in 1776, and sixteen hogsheads of juice arrived in 1778. It hardly mattered to the English who Jesse Fish was, and it didn't seem to matter to Jesse Fish who his customers were. Fish was a Yankee, a native of New York and by sympathy a revolutionary. Decades before the Revolution, he had retreated to an island off St. Augustine to get away from a miserable marriage, and he had become Florida's first orange baron.

There would have been others before him if Florida

had not been Spanish for more than two hundred years.
Andalusia and Valencia were golden with oranges, so the
Spaniards hardly needed to start commercial groves on
the far side of the ocean. They planted citrus fruit in remote places only for medical reasons. Columbus, under
orders, carried with him the seeds of the first citrus trees
to reach the New World, and he spread them through
the Antilles, where they grew and multiplied so vigorously that within thirty years some Caribbean islands
were covered with them. In all likelihood, Ponce de León
introduced oranges to the North American mainland
when he discovered Florida in 1513. (The Florida Citrus
Commission likes to promote him as a man who was
trying to find the Fountain of Youth but actually
brought it with him.) Hernando de Soto planted additional orange trees during his expedition to Florida in
1539. On all Spanish ships bound for America, in fact,
each sailor was required by Spanish law to carry one
hundred seeds with him, and later, because seeds tended
to dry out, Spanish ships were required to carry young
trees instead. Sir Francis Drake leveled the orange trees
of St. Augustine when he sacked the town in 1586, but
the stumps put out new shoots and eventually bore fruit
again. Nearly all were Bitter Oranges. Indians, carrying
them away from the Spanish colony, inadvertently scattered seeds in the Florida wilderness, and Sour Oranges
began to grow wild. (The first oranges in California
were planted around 1800 at a Spanish mission. The first
commercial grove in California was established by a

trapper from Kentucky in 1841, on the present site of the Southern Pacific railroad station in Los Angeles.)

England, which had acquired sovereignty over Florida in 1763, gave it back to the Spanish toward the end of the American Revolution, and the Florida citrus business did not increase significantly for nearly forty years, through the period that is known in Florida history as the Second Occupation. After the territory became a possession of the United States, in 1821, orange groves expanded rapidly in the St. Augustine area and along the St. Johns River south of Jacksonville. There was also some planting, in a minor way, on the Indian River. In 1834, two and a half million oranges were shipped north from St. Augustine, but on February 8, 1835, the temperature dropped as low as eleven degrees, salt water froze in the bays, and fifty-six hours of unremitting freeze killed nearly every orange tree in Florida.

One grove remained undamaged. It had been planted in 1830 by Douglas Dummett, a young man in his twenties, whose father had moved to Florida from New Haven, Connecticut, to establish a plantation of sugar cane. The site young Dummett had picked for his orange grove was on Merritt Island, between Cape Canaveral and the mainland, with the warm tidal waters of the Indian River on one side and of the Banana River on the other. His ground was high, the soil was rich in shell marl, and he had used wild Sour Orange trees as rootstocks, budding them with sweet oranges from a grove about fifty miles to the north. After 1835, orange growing in Florida was revived with buds from Dummett's trees. From the Dummett grove came the oranges of the Indian River, whose reputation soon spread so far that czars of Russia sent ships to fetch them.

The Indian River is not actually a river but a tidal lagoon, about two miles wide in most places and one hundred and twenty miles long, running between the Florida mainland and the Atlantic barrier beaches. Merritt Island is close to the Indian River's northern end. Dummett used to go out to meet trading schooners in a thirty-foot sailing canoe, its gunwales riding near the waterline under a load of oranges packed in barrels between layers of dried Spanish moss. He had made the canoe from a single log of cypress with the help of another orange grower, Captain Mills Olcott Burnham, a Vermonter who had moved to Florida for his health. Burnham had regained his strength so dramatically that he could lift two fifty-pound kegs, one in each hand, and hold them wide apart. After Congress passed the Armed Occupation Act—giving a hundred and sixty acres of Florida land to anyone who could fight off the Seminoles and hold his ground for seven years—Burnham, in 1842, led an unlikely group on an expedition south from Merritt Island, opening up the Indian River and establishing a community near the present site of Fort Pierce.

Florida was the only wilderness in the world that attracted middle-aged pioneers. The young ones were already on their way west toward California. The subtropics may actually have been fiercer than the plains, in

that both areas had hostile Indians but Florida alone had
its stupendous reptiles. Florida, even then, appealed to
aging doctors, retired brokers, and consumptives; exam-
ples of each of these categories went bravely down the
Indian River with Captain Burnham. So did Cobbett
the Cobbler, who is described in contemporary accounts
as having had a chalky face and a bright-red nose, which
acted as a kind of wet bulb for whiskey. Cobbett the
Cobbler could smell the stuff a mile away, and if a bottle
was open anywhere in the settlement he would charge
through the woods in its direction, yelling so primitively
that the others instinctively reached for their rifles.
Crazy Ned, a sailor who had once fallen from the top-
mast to the deck of a ship, was a beardless and irritable
man whose injuries unfortunately made him appear, with
each step he took, to be attempting to dive into the
ground. Ossian B. Hart, who later became governor of
Florida, was a violinist. When he played in the evening
for his fragile and patrician wife, a periphery of snouts
would appear beyond the veranda of his cabin. Hart was
Heifetz to the alligators of the Indian River, and the
settlement was safe when he was playing. Old Phil and
Young Phil Herman might have been twins, not only
because their features were similar but because Young
Phil looked old and Old Phil looked young. They were
actually father and son, and they had a peculiar sense
of hospitality. After inviting guests to dinner, Old Phil
would scatter crumbs under the table. In the middle of
the meal, he would tap his foot, and up through a hole

in the floor boards would come one of his trained snakes. While the snake writhed in and out among the legs of the guests, getting all the corn bread it could hold, Old Phil and Young Phil writhed with laughter in their chairs. Near U.S. 1, at Ankona, Florida, there is a memorial tablet that was set up in 1926 by the citizens of St. Lucie County, where most Indian River oranges now come from. "This monument was erected to commemorate the first white settlement on Indian River," it says. "In toil and perils they laid the foundation for the safety we enjoy today."

The settlement might never have ended if Burnham had not been away on a trip when his family and all other families departed in terror of an Indian uprising. The Indians liked Burnham so much that they hung around his house and passed the time of day with him when he was there, helping him with his work and cooking their meals in his pots and pans, but Burnham was in Charleston selling turtles to English importers when the Seminoles killed the settlement storekeeper, who had been cheating them. Word spread that the Indians were massing for a general slaughter, and the settlers prepared to escape, by sea.

Indians were running up the beach as the escape boat pulled away. Major Russell, one of the settlement's leaders, was standing up in the boat like Washington crossing the Delaware. The Indians hated Russell and always had. One of them fired at him and nicked him in the arm. Feeling pain that night, Russell went into the boat's cabin and groped in the dark for a bottle of salve. Picking up a bottle of black ink by mistake, he poured it over his arm. When the sun came up, he thought he had gangrene. The others knew that it was ink, but they thought even less of Russell than the Indians did, and they said nothing.

None of them could have imagined what would happen. After the group reached St. Augustine and dispersed, Russell went to a doctor and told him to cut off his arm. The doctor said that the arm would get better, but Russell kept insisting, until the doctor did it. Russell died in Orlando thirty-one years later, presumably unaware of the actual truth about his gangrene. Captain Burnham, who had missed the evacuation of the settlement, happened to encounter his family in St. Augustine on his way home from Charleston. He took them back to the Indian River, but only as far south as Merritt Island, where he developed his orange grove and also became, for the rest of his life, the keeper of the lighthouse at Cape Canaveral.

After the Civil War, the grove of Douglas Dummett became increasingly celebrated. Not only was it the oldest in the state, but it also continued to be the largest. In New York, Dummett oranges were worth one dollar more per box than oranges from any other grove. Dummett didn't work particularly hard to achieve this kingship. Most of the time he was fishing, or hunting wildcats on the mainland with his pack of dogs. He was the fastest canoeman on the Indian River, and in the Seminole

Indian War he had led a company of the militia so courageously that he was remembered for it. For a number of years he was a member of the Florida House of Representatives. At home, he lived in a small log cabin, and his children and their mother lived in another cabin two hundred feet away. Dummett ate alone out in the grove or in the cabin. His wife had left him years earlier, and the woman who had mothered his son and three daughters was a Negro. The son, Charles, shot himself when he was sixteen. His death was called an accident, but some people on Merritt Island thought he had done it because of the shame he was made to feel for having Negro blood. Dummett himself died in 1873 and is buried in an unmarked grave in an Indian River grove called Fairyland.

New orange growers who arrived after the Civil War tended at first to settle north of the Indian River region along the St. Johns River. The Great Freeze of 1835 had been dismissed as a fluke. It was thought that the freeze had been caused by a mountainous iceberg lying somewhere off St. Augustine, and that another one was unlikely to come. The growers, for economic reasons, wanted to be near the port of Jacksonville. One of the newcomers was Harriet Beecher Stowe, who, with her husband, Professor Calvin Stowe, bought about thirty acres in the village of Mandarin, in 1868. Florida accepted her warmly, partly because she was quoted in the Northern press, soon after her arrival, as having said that people in Florida were "no more inclined to resist the laws or foster the spirit of rebellion" than people in a state like Vermont. An article in the St. Augustine *Examiner* expressed satisfaction "that Mrs. Stowe has done this little to repair the world of evil for which she is responsible in the production of *Uncle Tom's Cabin*." She taught Sunday school for Negro children in Mandarin, and taught them to read as well, and for seventeen years she ran a successful orange grove—apparently with very little help from her husband. The Professor had a white, airy beard that started on the crown of his head, made an island of his face, and stuck out a foot from his chin. He wore a small skullcap of flaming crimson, and he spent nearly all of his time reading books on the veranda of the Stowe house overlooking the St. Johns. "His red skullcap served mariners as a sort of daytime lighthouse," Mrs. Stowe wrote in an article in the *Christian Union*. In Northern markets, there was considerable demand for fruit boxes stenciled with the words "OR-ANGES FROM HARRIET BEECHER STOWE—MANDARIN, FLA." Mandarin had been called San Antonio under the Spanish, but the Americans had, more appropriately, renamed it, for a citrus fruit. The word "mandarin" is thought by many people to be a synonym for "tangerine." But tangerines are actually only a variety of mandarin that happened to originate in Tangier. All mandarins—including the Satsuma, the Clementine, the Cleopatra, the Emperor, and the King, or King Mandarin, Orange—have the so-called zipper skin that grows around the segments of the fruit like a loose-fitting glove.

The plantation society of the St. Johns was fairly metropolitan in contrast to life on the Indian River. Families had settled all along the Indian River, but even twenty years after the Civil War they were few enough so that when they saw a sail miles away they could usually tell by the cut of it who was approaching. At night, a family would go out in a small boat, light a lantern, talk, drift, and in thirty minutes catch enough fish to feed them for a week. On trips for supplies in sailboats, they would sometimes see ahead of them a darkness formed on the water's surface by five hundred acres of ducks. As a boat approached, the ducks would rise with a sound of rolling thunder, leaving on the water five hundred acres of down. Everyone slept on down pillows and down mattresses. The river was full of oysters. The shores were full of cabbage palms, whose hearts, boiled, were delicious. Currency was almost unknown. The nearest bank was in Jacksonville. When families put up Northerners who came for part of the winter, payment was often made by check at the end of a visit. For months, these checks would go up and down the Indian River as currency, until they had so many endorsements on them that they looked like petitions. In Titusville, near Merritt Island at the north end of the river, there was a group called The Sons of Rest. Any member who was seen with perspiration on his face was fined twenty-five cents. At the end of each month, the money was used to buy a pair of overalls for the member who had worked the least. A man named Cuddyback won four pairs of overalls in a row and the organization disbanded. There was one lawyer on the river. He raised oranges because his practice was so small.

In 1881, Hamilton Disston, a maker of saws from Philadelphia, bought four million acres from the State of Florida in tracts that went almost from coast to coast. He sold off two million acres to a British land company and smaller amounts to other land companies in the United States, keeping some to promote on his own. Tantalizing propaganda began to come out of the state. The early circulars that reached Northern cities and farms usually told the approximate truth, saying, for example, that "many men on the Indian River live entirely upon the returns of a few large trees, spending the whole year in hunting and fishing—doing no work." In the *Horticulturist*, there had been a straight-faced report that an expert named Al Fresco, who had eaten oranges from Europe, the Azores, the West Indies, Australia, and Melanesia, had found none to compare with the oranges of Florida. After the Disston purchase, new stories like that one came out of the state every day. Pamphlets said that if a person put about eighteen hundred dollars into a new grove, he could expect that it would soon be worth thirty thousand. New England farmers read these things and hurried to Florida to share the new paradise with farmers from Georgia and the Carolinas, against whom they had been fighting less than twenty years before. "There is nothing to prevent the establishment in Florida of a race of rich men who will rank with the plantation princes of the old South," the Atlanta *Constitution* decided. Land was ceded to railroads, and hundreds of miles of track were put down within a few years—reaching from Jacksonville south to Orlando and west to Gainesville, and continuing down the spine of Florida to open up the Ridge. "Orange growing," wrote a promoter-in-residence at the Silver Springs Land Company, "is no dead level of monotonous exertion, but one that affords scope for the development of an ingenious mind." Englishmen in particular found this sort of argument irresistible, and one of the curiosities of the orange fever of the eighteen-eighties is that a high proportion of the people it attracted were English. A writer named Iza Duffus Hardy noted that they were "bronzed, hearty, healthy-looking young fellows, high-booted, broad-hatted, with their cheery English voices and jovial laughs, who ride over—sometimes on half-broken Texan ponies —from their respective 'places,' many a mile away, to spend a social hour in town, and report their progress for the benefit and encouragement of those who have not yet 'settled.' This one a year or two ago was a doctor in London, this an artist, that a barrister."

The Dummett grove on Merritt Island became the setting for an absurd charade. Dummett had died in 1873, and in 1881 his place was bought by a fake Italian duke and a fake Italian duchess. He called himself the Duc di Castellucio, and he hurled himself into the orange business. His wife contributed prose to the Jacksonville *Union* and the Titusville *Florida Star*. "The Neapolitans

say 'Naples is a bit of paradise fallen from above,' "
wrote the Duchess. "If this saying is true, certainly a bit
of the same paradise has fallen on Indian River." The
Duke and Duchess built an octagonal wooden palace
that is still standing. Nearly all the rooms were octagonal,
too, but some of them became irregular hexagons after
the Duke and Duchess quarreled so bitterly that they had
a partition built precisely through the middle of the
house and never spoke to one another again. While they
were fighting, their oranges were flourishing, and the
Dummett grove remained the largest in the state. The
origins of the Duke remain obscure, but the Duchess was
Jenny Anheuser, daughter of the St. Louis brewer.

Out on the river, meanwhile, stern-wheelers called the
Ina, the Ibis, and the Indian River had begun to operate
on regular routes, carrying hundreds of tourists and
more settlers on every trip. Rockledge, just south of
Cocoa, was for a time the wealthiest winter resort in the
United States, and the travel writer C. Vickerstaff Hine
called the Indian River "this occidental Adriatic." More
and bigger stern-wheelers began to crowd its channels,
until, in 1893, they suddenly became obsolete. Henry M.
Flagler's Florida East Coast Railway had reached the
river, and soon paralleled the length of it on the way
to Miami.

To orange growers who had chosen reasonably good
land, it appeared that almost anything any promoter
had ever written was true. No one worried much about
freezes. For one thing, it was an era of scientific advances

in which triumph over nature seemed not only possible but inevitable. A cannon had been fired in the streets of Jacksonville in 1888 in the belief that the concussion would kill all the yellow-fever microbes in the air of the city. Then came the most destructive freeze of the nineteenth century in Florida. The Great Freeze of 1895 actually happened in two stages—a crippling one in December, 1894, then a killing one on February 8, 1895, that sent freezing temperatures all the way to the Florida Keys. Tens of thousands of trees were killed to the bud union, and thousands more were killed to the ground. More than a billion oranges had been shipped out of Florida in 1894. The freeze reduced that figure the following year by ninety-seven per cent.

Many immigrant growers went back to Europe, and American citizens left the state. Some of those who stayed sold palmetto fronds to European buyers for conversion into artificial palm trees. Others planted vegetables in the middles in their groves while they waited for new scions and new trees to grow. The state's orange crop regained its 1894 level in 1910, after more groves had been planted on the Ridge. The northern plantations of the nineteenth century, in the area of Jacksonville, were permanently abandoned. The population of Orlando was ten thousand in 1890 and two thousand in 1900. Not until the twenties did it reach ten thousand again.

The Dummett grove survived the 1895 freeze. Most of its trees have since been replaced or rebudded, but six of the original ones that were set out by Douglas Dummett in 1830 are still alive. Each has the girth of a middle linebacker on a professional football team. The reputation Indian River oranges established in the nineteenth century has never flagged. In the eighteen-nineties, *Blackwood's Edinburgh Magazine* said, "The Indian River orange is not to be mentioned in the same breath with ordinary oranges. It is a delicacy by itself, hitherto unknown in the world, and which Spain need never attempt to rival." By the nineteen-twenties, the term "Indian River" had taken on such a ring of unquestioned quality that cities seventy-five miles inland apparently decided they were seaports. The Ridge, for a while, became the west bank of the Indian River, and the words "Indian River" appeared on orange crates going out of all parts of Florida. This eventually led to a cease-and-desist order issued in 1930 by the Federal Trade Commission and to the formation, the following year, of the Indian River Citrus League, a growers' organization dedicated to keeping the fame of Indian River citrus unimpeachable and the name parochial. An official Indian River area was established in 1941. It begins about ten miles north of Daytona Beach and continues south through Titusville, Cocoa, Melbourne, Vero Beach, Fort Pierce, and Hobe Sound to Palm Beach. The western demarcation line runs along about fifteen miles inland, more in the south.

Over the past decade or so, the River has grown about one-tenth the number of oranges grown on the Ridge.

Each Indian River orange shipped North carries the words "Indian River" on its skin, and this has helped to foster the mistaken belief that the words signify a distinct type of orange rather than the area from which it comes. The same varieties are grown on the River as elsewhere—about fifty per cent Valencias, thirty per cent Pineapples, the rest Hamlins, Parson Browns, and so on. But Indian River oranges have about twenty-five per cent more sugar in them than oranges grown on the Ridge, and they contain more juice as well.

For a long time, people believed that salt airs coming off the Gulf Stream were somehow responsible for the quality of the Indian River orange, and many older growers still feel and will express a mysterious debt to the sea. Pomologists say that the salt-air theory is nonsense. Indian River soil is "heavy," as Floridians put it. That means, as one grower said to me, that "if you let it run through your fingers you can actually get your hands a little dirty." There is no doubt that it is richer than the deep sand on the Ridge; it holds nutrients and moisture better, and it grows a better tree. But the main reason Indian River oranges can be so good is that, until recently at least, most of them have been grown on Sour Orange rootstock. Sour Orange does poorly in the deep sands of the Ridge, and the vigorously foraging Rough Lemon is used there. But Sour Orange does well in the soil near the river. Experiments in Indian River groves have shown that a tree on Sour Orange rootstock will not produce as much fruit as a tree next to it that has been planted on Rough Lemon, but the fruit it does produce will be sweeter and juicier.

To tell the truth, I think the interior oranges, as Ridge oranges are sometimes called, have a little more spirit than oranges of the Indian River. I remember the first Indian River orange I ate when I went over there for a few days after several weeks on the Ridge. It was so sweet that it seemed just to melt away. It reminded me of a description I had read when I was hunting around in a ten-pound tome called *Memoirs of the Faculty of Science and Agriculture, Taihoku Imperial University*. Part of this volume summarizes the first book ever written on citrus fruit. Its author was a northerner who had gone to the south of China and had later rhapsodized in his book about the celebrated Milk Orange of Wen-Chou. "When it is opened, a fragrant mist enchants the people," he wrote. "It is called the Ju Kan, or Milk Orange, because of its resemblance to the taste of cream."

Most tourists who go to Florida probably have no idea what or where the Indian River is, but when they stop to buy oranges they often ask for Indian River fruit. As a result, roadside stands all over the state have huge billboards on their roofs which say "INDIAN RIVER ORANGES." Nearly all of these stands are legitimate. They have trucked-in fruit of the Indian River to satisfy the tourist demand. I remember a stand on the Ridge with a billboard that said, "PULITZER GROVES—PRIZE INDIAN RIVER CITRUS."

The ubiquitous appearances of the term have sometimes misled even regular winter-residents, some of whom think that "Indian River" signifies a type of orange rather than an area from which oranges come. One day in a restaurant in a central Florida town, I got into a conversation with a man who said he had spent the last twelve winters in Naples. He told me that few oranges were grown down there, but they were as fine as any grown anywhere because they were Indian River oranges. Naples is two hundred miles from the Indian River.

Tourists who stop at roadside stands and order a shipment of Indian River oranges almost always get what they order. The stands are merely showcases, and much of the fruit they show is plastic. The order slips are mailed to a packinghouse on the Indian River. For that matter, if a tourist orders Ridge fruit from a stand on the Ridge, the slip is mailed to a packinghouse. Two packinghouses take care of most orders from roadside stands—and department stores—in the state. A couple of years ago, the Indian River Citrus League found a number of roadside stands selling oranges from the Ridge across the counter in bags labeled "*Indian River.*" Not all these offenders were in remote parts of the state. One stand selling counterfeit Indian River oranges was on the east side of U.S. 1. A car veering into this stand might have knocked it into the Indian River.

On Merritt Island, I met an orange grower named Robert Hill, who showed me a good many trees in his grove that had lived through the 1895 freeze. He said that they had been set out by his grandfather, and that other trees on his property had been set out by his father. Narrow roads wind through Merritt Island between high walls of orange trees, which are interspersed with numerous houses of growers. Their holdings can be as small as five acres. Robert Hill's groves cover about fifty acres, and his house, one of the oldest on the island, is on a kind of high bank, perhaps thirty feet above the Indian River. Merritt Island was once covered with pines that yielded such fine lumber that it was known as Merritt Island mahogany. The pines still fringe the island in places, and they rise eighty feet or so above Hill's house. Pine needles cover the ground from the house to the water, and the trunks of the pines are two and a half feet thick. Among them, on Hill's property, is a high Indian mound. Through the pines, the river looks more like an estuary in Maine than a tidal lagoon in Florida. On the day that I was there, a steady breeze was blowing through the trees, and it was not difficult to see what had brought Douglas Dummett, Captain Burnham, and Robert Hill's grandfather to the Indian River.

Hill and I had lunch that day in a place called Ramon's, in Cocoa Beach, and as we were driving there, he pointed out to me the gantries of Cape Kennedy across a few miles of open sand. Ramon's was the sort of place that the eyes take two or three minutes to adjust to. It was as cool as it was dark, despite the smoke and the crowd, which included a high proportion of unattached women. Several bartenders appeared to be tiring under the strain of their work, although it was a few minutes before noon. A number of men were wearing sports shirts, but most were in business suits. Hill, in work clothes, seemed a little incongruous to me in Ramon's. He is a short, colloquial, and thoroughly unselfconscious man, and while we were drinking our second round of bourbon-and-water, he told me that places like Ramon's were what Merritt Island had needed for a long time. "This is living," he said.

The National Aeronautics and Space Administration has acquired three thousand acres of Merritt Island citrus groves. The trees have been leased back to their former owners, but the growers are not permitted to live in the groves or even to store equipment there. Hill's grove is part of the six thousand acres of citrus that remain outside of the space reservation. Aerospace industries, residential housing, and places like Ramon's are taking over so rapidly that in a few years there will be no citrus trees on the island, with the exception of those owned by the National Aeronautics and Space Administration. Hill's son has built a new house next to his father's, and he will probably stay there even if the family grove is cut down. He works for R.C.A. No space-age Chekhov is going to write a play called *The Orange Grove* about the Hill family of Merritt Island.

With aerospace and realty interests preëmpting the northern shores of the Indian River, new plantings have been made in the south. Until 1959, all groves were within two or three miles of the ocean. They had to be, since most of the length of the river is paralleled, a couple of miles inland, by vast savannas, which are largely under water nine months of the year. Much of the west bank of the river is a kind of loaf of ground, described by Floridians as a "bluff" even when it rises only thirteen feet above sea level. But it is high ground indeed compared to what lies beyond it. The savannas reach out to the western horizon, low and flat, filled with saw grass and cat-o'-nine-tails, small cypress trees, and occasional hammocks covered with cabbage palms. Otter live in the savannas, and alligators, wildcat, quail, deer, rabbit, wild turkey, water moccasins, and rattlesnakes. In 1959, the Minute Maid Company went into the savannas with earth-moving machines and heaved up a great ten-foot wall of earth surrounding seven thousand acres of marsh. Then they pumped out the water, graded the sandy soil, and planted six hundred thousand orange trees. It was an impressive feat, and it emboldened many other companies and syndicates to do the same.

Minute Maid itself has since made an even larger reclamation, a mere part of which, not yet fully bearing, is the largest lemon grove in the world. Lemons are even more sensitive to cold than oranges, and the 1895 freeze killed all the lemon trees in Florida. The new Minute Maid grove, which appears to be far enough south to be safe, marks their return—in any appreciable quantity —to the state.

I went out into the savannas one day with Hugh

Whelchel, the county agent of St. Lucie County, whose job is to be a kind of intelligence service, communicating to farmers and fruit growers information that derives from scientific and academic circles. Whelchel, a tall and bald man of about forty, was a citrus major in the Class of 1949 at Florida Southern College, and, like all Florida county agents, he is officially ranked as an associate professor of the University of Florida. Riding out into the marsh country with him, in his pickup truck, I asked him about the ten-inch laceless boots he was wearing, because I had noticed that nearly everyone who works in Florida orange groves wears them. "They keep the sand out," he said. "I like a loose-fitting boot like these better than a tight boot, because fangs can go through these and not get to your leg."

"How often do you see a rattlesnake?" I asked.

"Oh, hell, I guess I kill ten or fifteen a year."

"You keep a gun here in the truck?"

"No. I got a jack handle in the back there."

We drove up a hill of dirt, over a dike, and down into the Minute Maid reclamation, where young trees had just come into bearing for the first time. The perfect geometry of the groves on the Ridge, with straight middles stretching for hundreds of yards, had always seemed remarkable to me. Some of the middles in this grove ran on for five miles.

The ground had been engineered and sculpted so that rainwater would run into furrows in the middles, then through swales and ditches into a perimeter canal, running all the way around the grove, inside the dike. The water is pumped over the dike into state flood-control canals. Whelchel said that water management is the chief concern of all citrus growers of the Indian River area, whether in the new reclamation land or in the older groves. The water table is often less than three feet down, and that is as far as roots can go. Sometimes, in a single day, enough rain falls to saturate the soil, and growers have to be prepared to drain off the excess or they will lose their trees. In spring, however, it is so dry that they have to pump the water back into the groves or the trees will die for lack of it. To be fully prepared for the dry season, Minute Maid left one square mile of savanna, in the center of each of its vast groves, untouched. The square mile acts as a reservoir. Since a tree's branch structure is proportionate in size to its root structure, Indian River trees are set in raised beds, as they have been since the nineteenth century. In this way, more root structure can grow above the water table. Nonetheless, Indian River trees are smaller than trees of similar age on the Ridge.

Nearly all of the new orange plantings in the reclaimed savannas are on Rough Lemon rootstock. The trees grow faster, bear more fruit, and are less susceptible to virus diseases than they would be on Sour Orange rootstock. Each individual orange is not up to the usual standard of the Indian River, but that hardly matters, since most oranges are now grown for concentrate plants —for the frozen people, as the makers of concentrate

are often called in Florida—and a grower's profits are
determined more by volume than by quality. The Rough
Lemon rootstocks of the savanna plantations are disturb-
ing to Whelchel and to other county agents and to many
growers in the area. They feel that the rise of concentrate may be causing the end, in a sense, of the Indian
River. The savanna plantations are officially part of the
Indian River, and their owners could take advantage of
the name if prices were high enough for them to want
to market fresh oranges. Nearly all Indian River trees
used to be on Sour Orange, but fifty per cent of them
are now on Rough Lemon. "Packinghouses can be expected to discriminate against Rough Lemon fruit,"
Whelchel said. "But there is no guarantee of this. We
can lose the golden name of the Indian River if we start
shipping fresh fruit grown on Rough Lemon rootstock.
A smart fresh-fruit advertiser would plug oranges grown
strictly on Sour Orange."

One day, I went into an Indian River packinghouse to
watch the objects of all this concern being readied for
market. Citrus packinghouses are much the same wherever they are. In a sense, they are more like beauty par-
lors than processing plants. To make their oranges
marketably orange, packers can do two things, one of
which is, loosely speaking, natural and the other wholly
artificial. The first is a process that was once known as
"gassing," but the unpleasant connotations of that word
have caused it to be generally suppressed, and most people now say "de-greening" instead. Green or partly

green oranges are put into chambers where, for as much as four days, ethylene gas is circulated among them. The gas helps eliminate the chlorophyll in the flavedo, or outer skin, which is, in a sense, tiled with cells that contain both orange and green pigments. The orange ones are carotenoids, the green ones are chlorophylls, and the chlorophylls are so much more intense that, while they are there, the orange color will not show through. Both of these pigments are floating around in a clear, colorless enzyme called chlorophyllase, which will destroy chlorophyll on contact but has no effect on anything else. The chlorophyll is protected from the enzyme by a thin membrane called a tonoplast. In chilly weather, the tonoplast loses its strength and breaks down, and the enzyme gets at the chlorophyll and destroys it. The orange becomes orange. It would seem to be simple enough to pick a green orange and put it in a refrigerator until it turns orange, but, unfortunately, the membrane that protects the chlorophyll from the enzyme will no longer react in the same way once an orange is picked. In the early years of this century, Californians noticed that oranges tended to become more orange in rooms where kerosene stoves were burning. Assuming that the heat was responsible, orange men on both coasts erected vast wooden ovens called "sweat rooms," installed banks of kerosene stoves, and turned out vividly colored petroleum-smoked oranges. The more enthusiastic they got, the more stoves they put in. The emerging oranges were half dehydrated. And, with alarming frequency, whole packinghouses would burst into flame. That era closed when it was discovered that the ethylene gas produced in the combustion of the kerosene was the actual agent that was affecting the color of the oranges. Ethylene appears to anesthetize, or at least to relax, the membrane that protects the chlorophyll. All fruits take in oxygen and give off carbon dioxide through their skins, and some fruits, interestingly enough, give off ethylene gas as well when they breathe. A pile of green oranges will turn color if stored in a room with enough bananas. One McIntosh apple, puffing hard, can turn out enough ethylene to de-green a dozen oranges in a day or two.

Oranges are gassed in both California and Florida, often merely to improve an already good color. A once orange orange which has turned green again on the tree will not react to it. Neither will an orange that is not ripe.

The second method of affecting the color of oranges is more direct: they can be bathed, at times, in a dye whose chemical name is 1-(2, 5-dimethoxy-phenylazo)-2 napthol, popularly known as Citrus Red No. 2. This is the only dye permitted by federal law. The use of it is against California law, and the law of Florida, as a kind of safeguard against criticism, requires that dyed oranges be labeled as such and that they contain ten per cent more juice than the established minimum for undyed oranges. In practice, this regulation affects only the Ridge, because Indian River packinghouses are, almost without exception, too proud to dye their oranges. Citrus Red No. 2 is an aggressive and unnerving pink, but, applied to the green and yellow-green and yellow-orange surfaces of oranges, it produces an acceptable color. How acceptable seems to differ with individuals, and, in a more remarkable way, with geography. Judging by sales figures, people in New England instinctively reject oranges that have the purple letters "COLOR ADDED" stamped on their skins. In the Middle West, though, color-added oranges are in demand. Stores have even put advertisements in Chicago newspapers announcing when color-added oranges were available. Distributors there say they could sell many more oranges if the packinghouses would intensify the dye. In Florida, citrus men sometimes say of Midwesterners, "These people don't want oranges. They want tomatoes."

In an average year, the color-add season only lasts for several weeks in the autumn. There is no need for dye in the winter. In the late spring, high-season Valencias may turn partly green again; the dye has an unsatisfactory effect on them, however, and is seldom used. In the course of a Florida season, as one variety follows another, there is a general rise in the internal quality of oranges, and high-season Valencias are, on the average, the best oranges available at any time in the year. They are often mottled with green, though, and many people pass them by.

The oranges in the Indian River packinghouse I visited had been gassed and were being washed in warm soapy water, brushed with palmetto-fiber brushes, and dried by

foam-rubber squeegees and jets of hot air. Brushed again
with nylon bristles to bring out their natural shine, they
were coated with Johnson's Wax until they glistened like
cats' eyes. The wax, which is edible, replaces a natural
wax that is lost when the oranges are cleaned. Apples,
cherries, and the rest of the pip and stone fruits look
much the same when they enter a packinghouse as when
they leave, but when oranges arrive they are covered
with various things, from sooty mold to dust smeared
by heavy dews. They have to be washed, but without
their surface wax they would breathe very rapidly and
begin to shrivel within hours. The natural wax therefore
has to be replaced. Packers replace it and then some—
but if they apply too much wax, the orange will suffo-
cate and its flavor will become, at best, insipid.

Oranges are graded after they are waxed. Eight ladies
stood beside a conveyor of rolling oranges, taking out
the ones whose skins were blemished by things like wind
scars, oil blotches from petroleum sprays, hail damage,
and excessive russeting on the blossom end. The oranges
they eliminated would go to concentrate plants. Elim-
inations, as they are called, are something quite different
from culls—split or rotting oranges that are taken out
before the beautifying process begins. The ladies who
do the grading wear gloves, because a light pass of a
fingernail over the surface of an orange can rupture oil
cells, causing peel oil to well out onto the surface and
not only discolor the orange but also nurture fungi that
destroy it.

Oranges do not bruise one another the way apples sometimes do, and an orange, in fact, can absorb a blow that would finish an apple. Oranges can actually be bounced like rubber balls without damage, except that when they are dropped more than a foot they will start to breathe too rapidly. Truck drivers who bring them into the packinghouses from the groves wade along through rivers of oranges while they are emptying their trucks, taking care to slide their feet along the bottoms of the chutes, like trout fishermen moving upstream. Government inspectors, who work every day all day in packinghouses and concentrate plants, squeeze a standard boxful—an average of about two hundred oranges—from each truckload, and they order the entire lot destroyed if the sample is not about ten per cent sugar and does not amount to at least four and a half gallons of juice.

Oranges that happen to be going to New York cross the Hudson River on barges and enter the city at Pier 28 at the western end of Canal Street. All fresh fruit of any kind that is shipped to New York City for auction is sold at Pier 28. The pier's interior is like the inside of an aircraft hangar, and fruits from everywhere are stacked in lots in long, close rows—oranges and grapefruit from the Ridge, California oranges, apples, avocados, pears, plums, cherries, lemons, grapes, pomegranates, and so on. Over at one side, separated by a wide area from all the other crates and boxes, is the fruit of the Indian River. A man from the Indian River

is always there to look after it, and he has no counter-
part elsewhere on the pier. Buyers walk around making
notes, then they go upstairs into a room that could
have been built as the auditorium of a nineteenth-century
high school. The walls are made of tongue-and-groove
boards and the wooden seats are set on frameworks of
cast iron, which are bolted to the floor. The room seems
to contain about ninety men and ninety lighted cigars.
In London in the eighteenth century, oranges were auctioned "by the candle." A pin was pushed through a
candle not far from the top, and when the candle was
lighted, the bidding began. When the pin dropped, the
most recent bidder got the oranges. In New York in the
present era, oranges appear to be auctioned by cigar.
The air in the auction room gets so heavy with smoke
that if anything as light as a pin were to drop, it would
probably stop falling before it reached the floor. The
auctioneer sits on a stage, usually alone. The man from
Indian River sits next to him when he auctions the fruit
of the Indian River.

SIX

DEGREES BRIX

On the Ridge as on the River, the organization of the Florida citrus business has acquired an elaborate superstructure; to the eye of the outsider it appears that for every man in blue jeans there are three with briefcases. The industry has an almost bizarre number of administrative and trade organizations, but there are five major ones—the Florida Citrus Mutual, the Florida Citrus Commission, the Florida Canners Association, the Florida Fresh Fruit Shippers Association, and the Indian River Citrus League. The offices of the Florida Citrus Mutual are in Lakeland, in a three-story building painted bright yellow with paint made experimentally from grapefruit. On an exterior wall, large lettering says, "IF IT'S GOOD FOR THE GROWER, LET'S DO IT," summing up the purpose of the organization inside. The Florida Citrus Mutual exists to create higher profits for grove owners, and it can exert considerable influence on the price of oranges through its market bulletin. Robert Rutledge, the organization's executive vice-president and operating head, is essentially a public-relations man. In newspapers like the Orlando *Sentinel,* the Lakeland *Ledger,* and the Tampa *Tribune,* his picture seems to appear more frequently than President Johnson's. He lobbies in Tallahassee, crowns citrus queens on the Ridge, and in general keeps himself and his organization so much in view that he has become a symbol of the modern Florida citrus business.

In the lobby of the Mutual building there is a picture, three feet by four, of Rutledge leaning against a ladder in an orange grove, wearing blue jeans and carrying a picker's bag. When I met him, one day, in his office, he was wearing a dark suit, a pastel-blue tie against a pastel-blue shirt, and black, laceless filigreed shoes. "I consider the whole state of Florida my home," he told me.

"Were you born in Florida?" I asked.

"I'm part Seminole."

On his briefcase were the words, in gold letters, "BOB RUTLEDGE, ACHIN' ACRES, FLORIDA." "Achin' Acres is the largest ten-acre ranch in the world," he said. "I have two thousand citrus trees on it. It's really seventy acres." At the time, he commuted to Achin' Acres in a shiny yellow Thunderbird with a grapefruit paint job.

"Have you always lived at Achin' Acres?"

"I was born in an alligator swamp," he said. The alligator swamp turned out to be a section of Peoria, Illinois. Rutledge went to Peoria High School and the University of Illinois, where he majored in marketing and was graduated in 1947. He was deeply tanned, and had dark, quick eyes that darted about the room as we talked. Beside the door of his office there was a plastic orange tree. On his desk were fourteen small brass plates, engraved with inscriptions like "Why All This Toil for the Triumphs of an Hour?," "Who Escapes a Duty Avoids a Gain," and "A Fool Flatters Himself, the Wise Man Flatters the Fool." On the wall was a framed copy of a poem called "My Florida":

> My Florida, when from thy low-hung stars
> Thy numerous inlets and thy tide-swept bars,
> I take my leave, and in the fading light
> My spirit into unknown parts takes flight . . .

One of the services that the Florida Citrus Mutual performs for the good of growers is a program of theft prevention. In some years, particularly after freezes, as much as two million dollars' worth of oranges and other citrus fruits have been stolen from Florida groves. After the 1962 freeze, Mutual set up a Central Intelligence Bureau. The central intelligence is Leslie Bessenger, a retired sheriff, who coördinates the efforts of sheriffs and deputies in thirty-two orange-growing counties. Bessenger is a heavy man with a leonine white head, and there is apparently nothing he would rather talk about than fruit thieves.

Sometimes thieves make a major haul, such as a whole semitrailer full of oranges; but the moonlighter, picking alone in the dark, is the most common kind. "The aggregate of the small steal is the major part of the big steal," Bessenger explains. One night a couple of seasons ago, a deputy in Bessenger's network noticed a white Cadillac whose underpinnings were all but scraping the road. The car was riding low enough to be a Chris-Craft. It had three thousand five hundred oranges in it, loaded to the windowline. The driver admitted that he was making his ninth haul in three weeks, and said that he could pick a Cadillacful of oranges in the dark in three hours.

Fishing boats anchor off orange groves during the afternoon and the occupants fish until dark; then they jump out of the boats with burlap bags and fan out into the groves. More ambitious thieves sometimes hire innocent pickers and clean out whole blocks of trees. "But the biggest deal I know of was at a processing plant," Bessenger said. "There was a seventy-one-thousand-dollar loss documented, but I believe it was well over a hundred thousand. The scalemaster was cheating—at a dollar seventy-five a box." This was the scalemaster's kickback for each box of nonexistent fruit he wrote up. Semitrailers roll into concentrate plants, one after another, all day long. Bessenger said that the crooked scalemaster was taking the plant for eight hundred dollars a truckload.

The Florida citrus industry is self-regulating and pays its own way. It has no land in the federal soil bank and receives no government subsidies. In a sense, the industry subsidizes government. The Florida Citrus Commission was set up by the state in 1935 to administer—among other things—the state's code of regulations for the citrus business. It is financed not by the state's general treasury but by a levy against growers.

Scientific research takes five per cent of the Citrus Commission's budget, and administration another five per cent. All the rest goes into advertising, promotion, market research. The commission has field men all over the United States feeding information to its economists in Lakeland, who bubble over with vital information: the more educated a person becomes, the more frozen concentrated orange juice he can be expected to buy; farmers, craftsmen, and laborers buy the least concentrate; office workers and sales people do better; doctors, dentists, lawyers, and corporate executives are the heaviest consumers of concentrate. Chilled single-strength juice—the kind that comes in milk containers in supermarkets—is preferred, typically, by the family of a highly educated middle-aged man who has two children and lives in New York. Sales of canned single-strength orange juice—the ancient kind of canned orange juice—went down sixty-five per cent between 1950 and 1965. There are no better consumers of canned single-strength juice today than the family of a blue-collar worker who has a grammar-school education, has several children under six, and lives in a Southern state.

There are twelve commissioners—all growers or canners or packers and nearly all millionaires—and they are appointed by the Governor of Florida. They meet in public session in a softly lighted, deeply upholstered, cool auditorium in the Citrus Commission's building in Lakeland. Their mission in recent years has been to sustain the miracle of concentrate. At one meeting that I attended, they appeared to be a worried group of men. Totally synthetic juice had suddenly come out of nowhere and had already taken fifteen per cent of the market. One synthetic juice was even being served regularly at the United States Naval Academy. There was talk of using deficit financing to launch an epic new advertising campaign, and debate grew tense over a proposal to attack the competition by adding sugar to frozen concentrate —until then such an unthinkable practice that anyone caught doing it had been fined fifty thousand dollars. Onstage, sitting behind a large U-shaped table, the commissioners had the manner of a high court, listening to testimony and arguing among themselves. At one point, the Commission's attorney, D. B. Kibler III (he is inevitably referred to among citrus people as Kibler the Quibbler), said that a point of law required that it be clearly established whether sugar could in any way be considered a preservative. Dr. Louis Gardner MacDowell was asked to take the stand.

A tall, fairly heavyset man wearing khaki trousers and a faded plum-colored sports shirt got up and walked down to the front of the auditorium. MacDowell is revered from Coral Gables to Pensacola Bay as the originator of concentrate. From his mind came the idea that produced a seven-hundred-million-dollar industry. However, as a public servant—he was, and still is, Director of Research at the Citrus Commission—he got nothing for his idea except six months of extra vacation time, which he has been using up by adding a week a year to his standard three. His salary is fifteen thousand dollars a year. He has often said, with equanimity, that if he had wanted to be a rich man he wouldn't have got a Ph.D. in the first place. He hunts when he can, fishes more often, and is kept fairly busy directing the research of forty-three men. One of the commissioners leaned forward and asked MacDowell whether, in his opinion, sugar could in any way be considered a preservative. "No," MacDowell said, and, without adding a syllable, he turned around and went back into the relative darkness of the auditorium.

The patron saint of concentrate, as he is sometimes called, was born in Asbury Park, New Jersey. He grew up mainly in Melbourne, Florida, on the Indian River. His father owned hotels in both towns. He remembers going up to Merritt Island as a boy and watching the orange boats being loaded from jetties. He was in the Class of 1933 at the University of Florida, in Gainesville. "I majored in chemistry," he told me, in the course of a conversation I had with him one day in his office. "But most of my time was spent boozing and hunting girls." He stayed in Gainesville to take his doctorate, and he became the seventh Ph.D. in the university's history. He was named Director of Research at the Citrus Commission in 1942, charged with finding new uses for oranges. He did not actually invent concentrate. Nor did he discover it. Various people working in the field had already found that it was easy enough to concentrate orange juice; the trouble was that it had almost no taste. When the water was removed, most of the orange flavor went with it. This remains true of concentrate today. When the evaporators are finished with the juice, it has a nice orange color and seems promising, but if it is reconstituted into "orange juice" it tastes like a glass of water with two teaspoons of sugar and one aspirin dissolved in it. It was this problem that MacDowell overcame. "Just as simple as that, it occurred to me to overconcentrate the juice and then add fresh juice to it," he said. "And that was that. Very simple. No earthshaking research at all." This process is known as "cutback" in concentrate plants. After the juice is concentrated to a super-thick viscosity, fresh orange juice and other flavoring elements are put into it. "It was my idea to do this, but my colleagues Dr. E. L. Moore and Mr. C. D. Atkins did the doing," MacDowell said. A patent was eventually granted to the three of them and assigned to the Secretary of Agriculture for free use by the people of the United States. The patent office had previously turned

down the application twice, because MacDowell's flash
of genius seemed to lack the stature of a discovery. After
two million gallons of concentrate had been sold, the
patent office changed its mind.

"I am a native of Florida, and I have a great love for
the state," MacDowell told me, with some irony coming
into his voice. "But Florida hasn't really got much of a
sense of romance about its citrus industry. California has
the original Navel Orange tree in a state park or somewhere. Nobody saves old trees here. Pink grapefruit developed in Bradenton. Parson Brown Oranges in Web-
ster. And so on. Nobody cares. We're the Fords and the
Chevrolets, not the Rolls-Royces and the Cadillacs. In
Indian River, they have some feeling that citrus is a way
of life. Over here, on the Ridge, citrus to most people
is just a way to make money."

The enormous factories that the frozen people have
built as a result of MacDowell's idea more closely resemble oil refineries than auto plants. The evaporators
are tall assemblages of looping pipes, quite similar to the
cat-cracking towers that turn crude oil into gasoline.
When oranges arrive, in semitrailers, they are poured
into giant bins, so that a plant can have a kind of reservoir to draw upon. At Minute Maid's plant in Auburndale, for example, forty bins hold four million oranges,
or enough to keep the plant going for half a day. From
samples analyzed by technicians who are employed by

the State of Florida, the plant manager knows what the juice, sugar, and acid content is of the fruit in each bin, and blends the oranges into the assembly line accordingly, always attempting to achieve as uniform a product as possible. An individual orange obviously means nothing in this process, and the rise of concentrate has brought about a basic change in the system by which oranges are sold.

Growers used to sell oranges as oranges. They now sell "pounds-solids," and modern citrus men seem to use the term in every other sentence they utter. The rise of concentrate has not only changed the landscape and the language; it has, in a sense, turned the orange inside out. Because the concentrate plants are making a product of which the preponderant ingredient is sugar, it is sugar that they buy as raw material. They pay for the number of pounds of solids that come dissolved in the juice in each truckload of oranges, and these solids are almost wholly sugars. Growers now worry more about the number of pounds of sugar they are producing per acre than the quality of the individual oranges on their trees. If the concentrate plants bought oranges by weight alone, growers could plant, say, Hamlins on Rough Lemon in light sand—a scion, rootstock, and soil combination that will produce extremely heavy yields of insipid and watery oranges.

As the fruit starts to move along a concentrate plant's assembly line, it is first culled. In what some citrus people remember as "the old fresh-fruit days," before the Second

World War, about forty per cent of all oranges grown in Florida were eliminated at packinghouses and dumped in fields. Florida milk tasted like orangeade. Now, with the exception of the split and rotten fruit, all of Florida's orange crop is used. Moving up a conveyor belt, oranges are scrubbed with detergent before they roll on into juicing machines. There are several kinds of juicing machines, and they are something to see. One is called the Brown Seven Hundred. Seven hundred oranges a minute go into it and are split and reamed on the same kind of rosettes that are in the centers of ordinary kitchen reamers. The rinds that come pelting out the bottom are integral halves, just like the rinds of oranges squeezed in a kitchen. Another machine is the Food Machinery Corporation's FMC In-line Extractor. It has a shining row of aluminum jaws, upper and lower, with shining aluminum teeth. When an orange tumbles in, the upper jaw comes crunching down on it while at the same time the orange is penetrated from below by a perforated steel tube. As the jaws crush the outside, the juice goes through the perforations in the tube and down into the plumbing of the concentrate plant. All in a second, the juice has been removed and the rind has been crushed and shredded beyond recognition.

From either machine, the juice flows on into a thing called the finisher, where seeds, rag, and pulp are removed. The finisher has a big stainless-steel screw that steadily drives the juice through a fine-mesh screen. From the finisher, it flows on into holding tanks. Orange juice squeezed at home should be consumed fairly soon after it is expressed, because air reacts with it and before long produces a bitter taste, and the juice has fatty constituents that can become rancid. In the extractors, the finishers, and the troughs of concentrate plants, a good bit of air gets into the juice. Bacilli and other organisms may have started growing in it. So the juice has to be pasteurized. In some plants, this occurs before it is concentrated. In others, pasteurization is part of the vacuum-evaporating process—for example, in the Minute Maid plant in Auburndale, which uses the Thermal Accelerated Short Time Evaporator (T.A.S.T.E.). A great, airy network of bright-red, looping tubes, the Short Time stands about fifty feet high. Old-style evaporators keep one load of juice within them for about an hour, gradually boiling the water out. In the Short Time, juice flows in at one end in a continuous stream and comes out the other end eight minutes later.

Specific gravity, figured according to a special scale for sugar solutions, is the measurement of concentrate. The special scale, worked out by a nineteenth-century German scientist named Adolf F. W. Brix, is read in "degrees Brix." Orange juice as it comes out of oranges is usually about twelve degrees Brix—that is, for every hundred pounds of water there are twelve pounds of sugar. In the Short Time, orange juice passes through seven stages. At each stage, there are sampling valves. The juice at the start is plain, straightforward orange juice but with a notable absence of pulp or juice vesicles.

By the third stage, the juice is up to nineteen degrees Brix and has the viscosity and heat of fairly thick hot chocolate. The flavor is rich and the aftertaste is clean. At the fifth stage, the juice is up to forty-six degrees Brix—already thicker than the ultimate product that goes into the six-ounce can—and it has the consistency of cough syrup, with a biting aftertaste. After the seventh stage, the orange juice can be as high as seventy degrees Brix. It is a deep apricot-orange in color. It is thick enough to chew, and its taste actually suggests apricot-flavored gum. Stirred into enough water to take it back to twelve degrees Brix, it tastes like nothing much but sweetened water.

As a season progresses, the sugar-acid ratio of oranges improves. Pineapple Oranges, at their peak, are better in this respect than Hamlins at theirs; and Valencias are the best of all. So the concentrators keep big drums of out-of-season concentrate in cold-storage rooms and blend them with in-season concentrates in order to achieve even more uniformity. Advertisements can be misleading, however, when they show four or five kinds of oranges and imply that each can of the advertiser's concentrate contains an exact blend of all of them. It would be all but impossible to achieve that. The blending phase of the process is at best only an educated stab at long-term uniformity, using whatever happens to be on hand in the cold rooms and the fresh-fruit bins. The blending is, moreover, merely a mixing of old and new concentrates, still at sixty degrees Brix and still all but tasteless if reconstituted with water.

The most important moment comes when the cutback is poured in, taking the super-concentrated juice down to forty-five degrees Brix, which MacDowell and his colleagues worked out as a suitable level, because three cans of tap water seemed to be enough to thaw the juice fairly quickly but not so much that the cooling effect of the cold concentrate would be lost in the reconstituted juice. Cutback is mainly fresh orange juice, but it contains additional flavor essences, peel oil, and pulp. Among the components that get boiled away in the evaporator are at least eight hydrocarbons, four esters, fifteen carbonyls, and sixteen kinds of alcohol. The chemistry of orange juice is so subtle and complicated that most identifications are tentative, and no one can guess which components form its taste, let alone in what proportion. Some of these essences are recovered in condensation chambers in the evaporators, and they are put back into the juice. The chief flavoring element in cutback is d-limonene, which is the main ingredient of peel oil. The oil cells in the skins of all citrus fruit are ninety per cent d-limonene. It is d-limonene that burns the lips of children sucking oranges. D-limonene reddened the lips of the ladies of the seventeenth-century French court, who bit into limes for the purpose. D-limonene is what makes the leaves of all orange and grapefruit trees smell like lemons when crushed in the hand. D-limonene is what the Martini drinker rubs on the rim of his glass and then drops into his drink in a twist of lemon. The modern Martini drinker has stouter taste buds than his predecessors of the seventeenth century, when people in Europe used to spray a little peel oil on the outside of their wineglasses, in the belief that it was so strong that it would penetrate the glass and impart a restrained flavor to the wine. In the same century, peel oil was widely used in Germany in the manufacture of "preservative plaguelozenges." In the fourteenth century in Ceylon, men who dived into lakes to search the bottom for precious stones first rubbed their bodies with orange-peel oil in order to repel crocodiles and poisonous snakes. Peel oil is flammable. Peel oil is the principal flavoring essence that frozen people put into concentrated orange juice in order to attempt to recover the flavor of fresh orange juice. "We have always had the flavor of fresh oranges to come up against," MacDowell told me. "People who make things like tomato juice and pineapple juice have not had this problem."

Because of freezes and other variables, concentrate has its good and bad years. In the past decade, for example, the '55s and '59s were outstanding. The '60s and '63s were quite poor. The '58s were even worse. But the '64s were memorable. Concentrate plants lay down samples in a kind of frozen reference library—one six-ounce can from each half hour of each day's run. The relative excellence of any given concentrate year is established by taste panels of citrus scientists, who stand in black-walled booths that are lighted by red light bulbs and drink concentrate from brandy snifters. They decide, variously, whether the taste is stale, insipid, immature, or overmature; too sour, too sweet, too bitter, or too astringent; whether it seems to have been overheated or to contain too much peel oil; and whether it suggests buttermilk, cardboard, castor oil, or tallow.

Plants that make "chilled juice" are set up as concentrate plants are, but without the evaporators. Instead, the juice goes into bottles and cartons and is shipped to places as distant as Nome. Tropicana, by far the biggest company in the chilled-juice business, ships twelve thousand quarts of orange juice to Nome each month. People in Los Angeles, surprisingly enough, drink two hundred and forty thousand quarts of Tropicana orange juice a month, and the company's Los Angeles sales are second only to sales in New York.

Tropicana used to ship orange juice by sea from Florida to New York in a glistening white tanker with seven hundred and thirty thousand gallons of juice slurping around in the hold. For guests of the company, the ship had four double staterooms and a gourmet chef. Among freeloaders, it was considered one of the seven wonders of commerce. To sailors of the merchant marine, it was the most attractive billet on the high seas. A typical week consisted of three nights in New York, two nights at sea, and two nights in Florida. There was almost no work to do. There were forty-two men in the crew, some with homes at each end. White as a yacht, the ship would glide impressively past Wall Street and under the bridges of the East River, put forth a stainless-steel tube, and quickly drain its cargo into tanks in Queens.

Tropicana unfortunately found that although this was a stylish way to transport orange juice, it was also uneconomical. The juice now goes by rail, already packed in bottles or cartons. The cartons are being phased out because they admit too much oxygen. Tropicana people are frank in appraisal of their product. "It's the closest thing to freshly squeezed orange juice you can get and not have to do the work yourself," one of the company's executives told me. To maintain the cloud in the juice and keep it from settling, enzymes have to be killed by raising the temperature of the juice to nearly two hundred degrees. Even so, there is some loss of Vitamin C if the juice remains unconsumed too long, just as there is a loss of Vitamin C if concentrate is mixed in advance and allowed to stand for some time.

During the winter, Tropicana freezes surplus orange juice in huge floes and stores it until summer, when it is cracked up, fed into an ice crusher, melted down, and shipped. In this way, the company avoids the more usual practice of chilled-juice shippers, who sell reconstituted concentrate in the summertime, adding dry juice-sacs in order to create the illusion of freshness. The juice-sacs come from California as "barreled washed pulp."

Leftover rinds, rag, pulp, and seeds at chilled-juice and concentrate plants have considerable value of their own. In most years, about fourteen million dollars are returned to the citrus industry through its by-products. Orange wine tastes like a one-for-one mixture of dry vermouth and sauternes. It varies from estate-bottled types like Pool's and Vino del Sol to Florida Fruit Bowl Orange Wine, the *vin ordinaire* of Florida shopping centers, made by National Grape Products of Jacksonville, and sold for ninety-nine cents. Florida winos are said to like the price. Florida Life cordials are made from citrus fruit, as are Consul gin, Surf Side gin, Five Flag gin, Fleet Street gin, and Consul vodka.

Peel oil has been used to make not only paint but varnish as well. It hardens rubber, too, but is more commonly used in perfumes and as a flavor essence for anything that is supposed to taste of orange, from candy to cake-mixes and soft drinks. Carvone, a synthetic spearmint oil which is used to flavor spearmint gum, is made from citrus peel oil. The Coca-Cola Company is one of the world's largest users of peel oil, as anyone knows who happens to have noticed the lemony smell of the d-limonene that clings to the inside of an empty Coke bottle.

A million and a half pounds of polyunsaturated citrus-seed oil is processed and sold each year, for cooking. Hydrogenated orange-seed oil is more like butter, by-products researchers told me, than oleomargarine. Noticing a refrigerator in their laboratory, I asked if they had some on hand. They said they were sorry, but all they had was real butter. Would I care for an English muffin?

Looking out a window over an orange grove, one researcher remarked, "We are growing chemicals now, not oranges." Dried juice vesicles, powdered and mixed with water, produce a thick and foamy solution which is used to fight forest fires. Albedecone, a pharmaceutical which stops leaks in blood vessels, is made from hesperidin, a substance in the peels of oranges. But the main use of the leftover rinds is cattle feed, either as molasses made from the peel sugars or as dried shredded meal. Citrus pulp and chopped rinds are dried for dairy feed much in the same way that clothes are dried in a home dryer —in a drum within a drum, whirling. The exhaust vapors perfume the countryside for miles around concentrate plants with a heavy aroma of oranges. The evaporators themselves are odorless. People often assume that they are smelling the making of orange juice when they are actually smelling cattle feed. If the aroma is not as delicate as the odor of blossoms, it is nonetheless superior to the aroma of a tire and rubber plant, a Limburger cheese factory, a pea cannery, a paper mill, or an oil refinery. Actually, the orange atmospheres of the Florida concentrate towns are quite agreeable, and, in my own subjective view, the only town in the United States which outdoes them in this respect is Hershey, Pennsylvania.

One plant is now concentrating juice electronically—and some citrus scientists think that this may be the process of the future. Into a tank of juice goes a large rod which gives off high-energy short-wave pulses, energizing water molecules in the juice. The energized particles leave more rapidly in the evaporator, and fewer evaporating stages are required. Electronic concentrate of seventy-two degrees Brix can be quickly produced and will keep at thirty-two degrees Fahrenheit, whereas ordinary concentrate requires temperatures around fifteen below zero.

Dr. MacDowell, and others, think that any form of concentrate may be just an intermediate phase. "This is what we're working on," he said to me when I visited him, reaching to a shelf for a laboratory display bottle which contained light yellow crystals of dry orange juice. "Ninety-nine Brix," he said. "These are the pounds-solids that everyone talks about and almost no one ever sees." Crystals are being made in several ways. MacDowell's team uses a process called foam-mat. They introduce very small amounts of methylated cellulose or modified soybean protein as foaming agents, then beat the juice until it is stiff, like egg whites. They spread this as a kind of orange-juice mat onto perforated steel trays and blow air up through the holes in the trays until the mat is dry. At Plant City, Florida, a company called Plant Industries, Inc., is already making orange-juice crystals commercially, and by a different method. They pour concentrate onto a moving steel belt in a hot vacuum chamber. The concentrate dries on the belt in eighty seconds and is scraped off. As it is packed, a bit of peel oil, locked into sugar granules, is sprinkled into each can.

Stirred into water, crystals taste like concentrate. Plant Industries sells them to hospitals and other institutions, and to the armed forces. The trouble with crystals commercially is that they are still too expensive. After all, the crystal-making process begins with finished concentrate. The general manager at Plant Industries is a Yale graduate who was raised in St. Paul, Minnesota. He says generously that crystals produced the other way, by the foam-mat process, make fine orange juice, except that it tends to get a head on it, like a glass of beer.

The Minute Maid Company was actually developed by a research organization in Boston which had been trying to make orange-juice crystals during the Second World War. When they decided to give up on crystals and market cutback concentrate instead, they built Florida's first concentrate plant. To the possible discomfort of some of his neighbors in California, Bing Crosby bought twenty thousand shares of Minute Maid and began to say mellifluous things on the radio about the Florida company that led the concentrate boom. Eventually Snow Crop concentrate was taken over by Minute Maid, and Minute Maid itself was swallowed by the Coca-Cola Company in 1960. Minute Maid has three concentrate plants in Florida and thirty thousand acres of Florida orange groves. Theirs is the great orange dukedom now. With the rise of concentrate, the era of individual orange barons is all but past. Minute Maid's central offices are in Orlando, in a nicely landscaped building that could be a suburban high school. When I first made a visit there, I called on Clifford Hodgson, chairman of the Minute Maid executive committee, who has since retired. I found him to be a warmly engaging and gracious man from Athens, Georgia, whose benign smile suggested that when he was a small boy he dreamed of having all the Coca-Cola he could ever drink, and that the dream came true. As a vice-president of Coca-Cola, he had been assigned to Minute Maid at the time of the merger. One morning, while he was telling me about the company's new eight-thousand-acre Hodgson Grove near the Indian River, a somewhat younger man named Ben Oehlert, Jr., walked in, briefcase in hand, saying that he just wanted to say hello and goodbye because he had to catch a plane. Oehlert, who had been working for Coca-Cola since 1938, was at that moment the president of Minute Maid, and he was later promoted to an even more commanding post in Coca-Cola headquarters in Atlanta. He was a tall man with a well-cared-for mustache, expensive-looking clothes, and a metropolitan manner. He said, "Never did I think as I grew up on the sidewalks of Philadelphia that I would end up as a farmer in Florida."

SEVEN

ORANGE BARON

THAT particular encounter, with the proconsuls of Minute Maid, caused me to wonder if anyone really powerful in citrus had been born in Florida, and whether the individual orange barons from the era that is referred to in Florida as the old fresh-fruit days had already vanished completely. Wherever I went, I asked people if they could think of a remaining example. In every instance, they immediately said, "Ben Hill Griffin." People told me, almost reverently, and with a sound of legend in their voices, that Ben Hill Griffin, of Frostproof, Florida, has his own personal concentrate plant, and that this is like having your own turnpike, or your own air force, or, at the very least, your own county. To give me an idea of the importance of Ben Hill—as everyone referred to him—people mentioned that he himself was too busy to take a direct part in the operation of the Florida Citrus Commission but that one of his office employees, a man named King Sprott, was at the moment serving as the Commission's chairman. Griffin, I gathered, is a universal man of citrus, a fresh-fruit shipper and a concentrate maker—one foot in the past and the other in the future. He is a major orange grower and a packer of every citrus product from old-fashioned, unfrozen canned orange juice to the pulp that is put into the orange drinks sold in Broadway theaters. On the side, he is both a cattle baron and a state senator, whose anti-segregationist convictions have been applied effectively in Tallahassee. People told me that the domains of Ben Hill are so endless that he has no idea how many acres of groves he owns. For that matter, they would add, Ben Hill's uncle, an aging citrus king named Lat Maxcy, has no idea how many acres *he* has, either. Ben Hill and Lat Maxcy have been involved for decades, I was informed, in a family feud that has become celebrated in the state, and they haven't spoken to one another for twenty years. Ben Hill has long since taken over much of what Maxcy once owned, but when Ben Hill started out, people told me, he slept under a truck; his Uncle Lat never gave him a thing, not even so much as one sick tree.

When I called Griffin to ask if I could come to see him, he said that the Senate was in session and he had to be in Tallahassee all week, but he would like me to join him and his family the following Saturday at his Peace River Ranch, in Hardee County, forty miles west of the Ridge, where he had some cattle. "We'll even give you a bed," he said in a spectacular drawl.

It was nearly dark when I got there, and Griffin had driven out to the gate because he thought I might miss the turn. A wiry man in his mid-fifties, of medium height, with thinning brown hair and light-blue eyes, he was wearing faded work clothes, with a red bandanna in his hip pocket. He had a slight forward lean, and he carried a long cigar in his hand. He got into my car. His own car, with four young men in it, led the way on into the ranch. He explained that they were his son and sons-in-law, and that one of them ran the ranch, one was the executive vice-president of a bank he owned, one ran the warehouse and shipping departments at his plant, and Ben Hill III was studying citrus at the Experiment Station. He said he would be acquiring one more son-in-law someday, and with this "team of five" he could face any dip in the economy by firing everybody else and keeping his captive lieutenants. It was a three-mile drive from the gate to his house.

His house, wide and low, looked like the main building of a working ranch, and reached miscellaneously to a tall stone fireplace and a porch seventy-two feet long, on which thirteen rocking chairs were set in a row. "I got me a jackleg carpenter and I built it myself," Griffin said. "That's a long time ago now. We started with ten acres here." A fire was going under an outdoor grill large enough to cook about fifty T-bone steaks. Before dinner, Griffin's wife asked me what I would like to drink, and when I said rum, Griffin poured a glass of rum made from oranges. The family ate at a long oak table. Above the table was the stuffed head of "Old Domeneker," a favorite bull, so named because his red-and-white spotted hide looked something like the plumage of a Dominique chicken. After he died, Old Domeneker spent one year in the cold room at the concentrate plant in Frostproof, because Griffin's taxidermist, who lives on the Gulf coast, was too busy with tarpon that had been caught by important tourists.

Later, Griffin sat down in a rocking chair on his porch and talked with me until three in the morning. He spoke in a remarkable mixture of colloquial and polished speech. When he talked about things like citrus prices and prognostications and the season's probable concentrate pack, he became very animated, waving his arms and speaking as if he were angry, pounding his fist on the rocker arm. He paused at one point to say that he has difficulty talking about anything he cares about without making people think he is getting ready to belt them. "Sometimes I may give the impression that I'm casual," he said, "but I work like hell. I get up every morning at six-thirty, and—"

"Do you drink orange juice with your breakfast?" I asked him.

"I drink orange juice six days a week," he said. "I drink tomato juice on Sunday to get my vitamins. I work like hell, as I was saying. I'm the hardest-working son of a bitch you ever saw. And I work everyone around me. I guess I work them too hard. I've never had any goal. I don't consider myself successful. The race isn't over yet. When I was a small boy playing marbles, I learned that the most important thing is position. If you get in the right position, you can clear up some marbles out of the ring." The air became damp and chilly, and he picked up a woman's cardigan that had been left on a chair next to him and set it over his shoulders as a kind of shawl. "The citrus industry is a killer of men," he said, without exclamation. "The captains of the industry die young—in their fifties or sixties. They work their brains and tissues too hard."

I asked him if he had any plans to acquire a second concentrate plant.

He said, slowly, "No man on earth ought to have more than one concentrate plant."

Griffin said that his father had been a citrus grower, with about a hundred acres of groves, and that his father always had some money, but not much. His father named him for Georgia's nineteenth-century Senator Benjamin Harvey Hill, a man who regarded the federal government as a caged monster and delivered eloquent speeches recommending that the key be thrown away. Griffin was raised in the groves—pruning, hoeing, fertilizing, running mule-drawn sprayers, climbing into a tree sometimes to get a shiner, only to find that a woodpecker had completely hollowed the shiner's interior. When he was a boy in Frostproof, the roads there had been paved with pine needles, spread in the ruts so that vehicles could move over the sand. "Frostproof should have been called Freezeproof to be literally true," Griffin said.

Ben Hill was in the Class of 1930 at Frostproof High School. "Football was my life," he said. "Basketball, too. I played second base on the baseball team." He only weighed a hundred and forty-five pounds, but he played football in the middle of the line, and during his three years on the team, almost no one scored on Frostproof. The score he seems to remember with the most relish was Frostproof 22, Sebring 0. He was too small to play for Florida. He went to the university for two and a half years, as a special student, studying agricultural economics and horticulture.

Once, when his father was unable to sell some grapefruit, Ben Hill borrowed a truck and took the grapefruit to Jacksonville. This was his first moment as a citrus baron. When he got there, he couldn't sell the grapefruit, had no money, and slept on straw under the truck. After a couple of days, he still had the grapefruit, but he was not going to admit defeat and take it home. He found a man with a load of cantaloupe and traded the grapefruit for it. No one wanted his cantaloupes. He traded them for a load of beans. Wiring home, he asked what the bean situation was in Frostproof. A telegram came back: NO BEANS IN FROSTPROOF. "I headed south with them beans," he remembers. Frostproof was indeed beanless, but apparently by choice. Almost no one wanted any of Ben Hill Griffin's beans. They began to fade, wilting and going soft. He dumped them on a lawn and sprinkled them with a hose, trying to bring them back. He found a few buyers, but had to throw most of the beans to hogs. "What the hell," he said, finishing the story, "I had nothing to lose."

Griffin's first wage-paying job was in a fresh-fruit packinghouse. "It was the L. Maxcy Company," he said. "Today I own it. It's part of Ben Hill Griffin, Incorporated." Maxcy paid Griffin twenty-five cents an hour, and Griffin remembers that he once got twenty-five dollars and twenty-five cents for a one-hundred-and-one-hour week—"and there was no deducts." When he was married, in 1933, his father gave him a ten-acre grove of Valencias, Pineapple Oranges, and five grapefruit trees. He bought a tractor and started looking after other people's groveland as well as his own. Over the years, nearly all his profits went into more groveland and more ranchland.

Griffin did not acquire the L. Maxcy Company directly, and probably never could have. Maxcy, who is now in his high seventies, sold his plant and groves to Snow Crop in the early nineteen-fifties, taking something like five million dollars, and establishing himself, almost permanently, on an outdoor bench near his own bank, the Citizens Bank of Frostproof, which is on Wall Street, in Frostproof. All Snow Crop really wanted was his groves, and the plant facilities deteriorated. Griffin offered more than a million dollars for them, and took over. "Lat Maxcy always knew exactly how much he was worth and how much land he had, and he still does," Griffin said. "And so do I." Griffin has about six thousand acres of citrus groves. The ranch on the Peace River is now fifteen thousand acres. He is worth about thirty million dollars.

In the morning, we went up in Griffin's helicopter, which was flown by Griffin's personal pilot. Five hundred feet over the ranch and its twelve buildings, Ben Hill pointed down to the airstrip that was built for his twin-engine Aztec, the twenty-acre duck pond, the full-size basketball court for the cowhands' children, the Allis-Chalmers road scraper, herd upon herd of Brahmin cattle, and wild bitter-orange trees with bitter oranges on them. Then we headed northeast, across scrub-pine land, toward Frostproof. Griffin seemed to become increasingly excited by the trip as we went along, despite the number of times he had flown it. When he spotted a couple of wild turkeys, he all but jumped out of the helicopter. Even at five hundred feet, the turkeys looked as big as small horses. Suddenly, the scrub-pine land came to an end. The rise of the earth below was not apparent from above, but we had come to the Ridge, and more than a hundred miles of citrus trees ran on and on and on through and beyond the haze at the northern horizon.

"There's where we started!" Griffin shouted, pointing toward a block of trees. "Right yonder, on that ten acres. I planted all of that next grove, too. Forty acres. I grew the rootstock in my back yard. I had a colored man lived on the original ten. Hilliard Baron. He had one hand,

and half of one foot was gone, but he cleared that second forty acres. Felled big oaks. If I had a hundred men like Hilliard, I could conquer the whole state of Florida." We flew over one Ben Hill Griffin grove after another. "Those are two-year-old trees right there," he said. "See how uniform they are. Every one is living. No skips." Up ahead, the Ridge was perceptibly higher, forming a vast low dome of earth, flat at the summit, where two enormous circular lakes almost touched one another. There was a town on the narrow strip of land between the two lakes. Griffin was really excited now. "There it is!" he shouted. "That's it! That is the Isthmus of Frost-proof!"

The air over the town was full of the aroma of Ben Hill Griffin's concentrate plant, as if the whole of Frost-proof was an orange cake, baking. He pointed out an automobile agency that he owns, a fertilizer plant that he owns, his Florida Spanish house, and an empty lot where Ben Hill III was about to start building a house. Around the two big lakes, Lake Clinch and Lake Reedy, dozens of smaller lakes, all of them round, were spaced out in every direction, and the earth, from the air, was all cir-cles of blue against a field of citrus green. The helicopter slid down into a town park behind the concentrate plant. Griffin got out and led the way, running, up a hill to a complex of buildings that included his concentrate plant and his fresh-fruit packinghouse. Outside was a flagpole, flying the flags of the United States and of Florida. He said he was having his own flag made and that it would soon be flying there, too. Running into the packing-house, he picked up an orange, took a knife out of his pocket, rapidly peeled a helical ribbon from the blossom end, and handed the orange to me. "When you've eaten the rest, you can use the peel for a handkerchief," he said. "You know, I'm just as lost as a by-god without a knife in my pocket. If you see me with my pants on, I've got a knife in my pocket. You can't tell what's in-side an orange from the outside. You've got to cut it." Moving on into the laboratory, where Ben Hill Griffin Concentrated Orange Juice is repeatedly tested, he picked up a beaker from the day's run, and handed it to me. At that moment, it seemed to me to taste like the fresh juice of a high-season Valencia, picked from far up on the south side of the tree.

Going back to the helicopter, we walked beside a rushing stream, which might have been a trout stream in Vermont, full of boulders, pools, eddies, and tumbling cascades. "That water is coming from the evaporator," Griffin said. "It was inside oranges a few minutes ago."